ALL ABOUT HOME ELECTRICS

ROY DAY

HAMLYN

London · New York · Sydney · Toronto

Contents

About the Author

Roy Day is an established authority on most aspects of design and maintenance in the home. His many television appearances and broadcasts have brought him before a wide public.

A lecturer of considerable experience in both interior design and the history of the subject, he is also a regular contributor to several monthly magazines.

He has written several books on do-it-yourself and is particularly interested in craft education for schools as well as for students of all ages. His special subject is the history of English and French furniture and architecture on which he has lectured for several years.

Primarily a designer, Roy Day is concerned with the practical aspects of design about the house, devoting as much time as possible to the actual construction of the many schemes which he originates. He is ever critical of the lack of attention which many designers of alleged repute show towards the problems of function and maintenance. Important though it is, mere visual appeal should never be an end in itself.

Many materials which can add so much to the attractiveness of a home require certain techniques and know-how to facilitate their use successfully and he feels that the visual demonstration and presentation of concisely illustrated books and magazines can do much to help the enthusiastic user to overcome recurring problems. He has been instrumental in persuading many manufacturers to devote more time and resources to instruct the consumer wisely on how to use their products to the best advantage.

Roy Day is also aware of the growing interest which women are showing towards do-it-yourself in the home, particularly in the craft subjects, and he is devoting more and more time to appearing in women's magazine programmes on television. He has borne the woman's interest in mind when writing this book.

Published by
The Hamlyn Publishing Group Ltd.
London · New York · Sydney · Toronto
Astronaut House, Feltham, Middlesex, England

ISBN 0 600 30299 7
Phototypeset by Tradespools Ltd, Frome, Somerset

Printed in England by
Chapel River Press, Andover, England

Acknowledgements
The publishers would like to thank MEM Co. Ltd., Maclamp Co. Ltd. and Burtt and Parker of Lightwater for their help with illustrations.
Line drawings by Stuart Perry

Introduction

Wiring up a plug and mending a fuse are basic electrical jobs most householders have to do. Although these jobs are fairly elementary, knowing how to do them correctly and safely is essential.

Sooner or later, also, some other parts of the domestic electrical system may need maintenance, or the need may arise for an extra light or a socket outlet. It is then that a basic grasp of how a house is wired becomes a valuable asset.

In this book I have tried to give enough guidance for straightforward wiring jobs to be carried out. The book is intended primarily as a guide for those who have little or no knowledge of electrics, not as a complete manual of wiring practice.

A word or two of warning. Electricity is a very dangerous thing. So don't attempt any rewiring job unless you are convinced you know exactly what you are doing.

Most important, and I repeat this warning several times in the book, always remember to turn off the current at the fuseboard or consumer unit before interfering with any wire, accessory or appliance which is connected to the system. Fires and accidents are caused every day by carelessness in handling electrical appliances. Make sure you do not become a victim through neglecting to take simple precautions.

Roy Day

Fig. 1 A selection of modern switches and sockets

Source of Supply

For simple wiring-up jobs and routine maintenance of a household's electrical system, it is usually sufficient to know what goes on on the consumer's (that is, your own) side of the installation. However, a basic understanding of the principles of electrical supply from the electricity company can be helpful.

Let us, then, take a brief look at the methods used to supply electricity to domestic premises without becoming involved in theory and too many technicalities.

SERVICE CABLES

Electricity is supplied from a power station to a house through two wires called conductors. One is red; the other is black. The red one is called the Live; the black is known as the Neutral. These wires are the service cables.

Current is supplied by the live wire. After it has entered the house, lighted the lamps and fed power to your various fires and other appliances, the current returns to the power station. This time it is the black (neutral) wire which carries the current.

The red wire is live at, usually, 240 volts. The black wire is connected to earth on its journey back to the power station.

The supply cable to the house leads into the electricity board's sealed fuses, through the meter which records the amount of electricity you use, to a main switch on your consumer unit or fuseboard.

The meter is situated near the consumer unit or fuseboard. This measures the electricity used in kilowatt hours.

A kilowatt is another way of saying 1,000 watts. If you use a 1,000 watt appliance for one hour, then you will consume 1 kilowatt hour of electricity in that period. One kilowatt hour (or 1 kWh as it is usually expressed) is the standard unit of electrical consumption.

THE COST

As your electricity bill will show, the cost of electricity used in a domestic property is calculated at so much a unit. If you know the cost of a unit and the wattage of an appliance, you can work out how much it costs to run the appliance, but you will need to be good at arithmetic to arrive at an accurate figure!

The reason for this is that electricity boards' charges per unit are rather complicated, especially since decimalisation. Prices, too, may vary from area to area. Some boards charge one figure for a certain number of units and another figure for the rest of the units consumed.

As an illustration, a bill I have before me reads:

Meter readings
Present, 40436
Previous, 39626
Units consumed, 810.

Of this total of units, 195 are charged at 1·52p. The rest (615) are charged at 1·27p. The total charge comes to £10.77 for current consumed. In addition, there is a standing charge of £1.05, so the total bill for the quarter is £11.82.

Most people, of course, will be only too familiar with electricity bills, but for the uninitiated, that is a general guide as to how they are made out.

It pays to keep a close eye on the amount you are charged for electricity as mistakes are made occasionally.

THE METER

However, if your bill should jump alarmingly in one quarter, don't immediately assume that the meter is inaccurate. As a matter of fact, they rarely go wrong.

It is unlikely, too, that the board's representative who reads the meter will make a mistake. If you were out when he called and supplied your own reading of the meter, a mistake could easily be yours. Later I will tell you how to read a meter.

So it is advisable, when a bill for electricity seems abnormally high, to check all possibilities before complaining to the board that you have been overcharged.

If you do suspect your meter is not working properly, turn off all your electrical appliances. Now look at the meter. Just below the dials is a circular disc. This is designed to revolve *only* when electricity is being consumed.

The speed at which the disc revolves depends on how many appliances are switched on at one time and upon their rating (how many kilowatts they consume).

An electric clock, for instance, uses hardly any electricity and you may find it difficult to detect a movement of the meter disc when a clock is the only appliance switched on in the house. But watch the meter disc and see the difference in the rate at which the disc turns when you switch on, say, a 2,000 watt fire or perhaps an immersion heater (which is usually 3,000 watts).

If, after turning off all your appliances (and lights too, of course) the meter disc is still revolving, there is something wrong. But before you call in the board to investigate, make absolutely sure an appliance has not been overlooked.

I emphasise this because I once heard of an instance where this sort of thing happened. After exhaustive checks had been made to find out why a meter disc continued to turn after all the appliances had been turned off, an electric heater was finally found switched on in the loft. It had been turned on at the beginning of a cold spell in the autumn and had burned away merrily until discovered the following spring!

Apart from the fact that this was misleading and indicated a fault, think of the waste of electricity! There could always be the danger of fire from an unseen heater as well.

CONSUMER UNIT

The meter is connected to the consumer unit by two wires (one red, the other black). These are called the meter tails and I shall refer to these again later on.

Apart from these wires, there is one

more wire – a very important one – involved in the installation. This is the earth or safety wire.

All installations must be earthed to comply with the regulations. This is done by linking all the installation's earth wires to a common earthing point at the consumer unit. This, in turn, is linked by the board to a special earthing terminal usually provided by the board.

In some areas it is not possible to bond the installation to a suitable point, so an earth leakage circuit breaker is fitted instead. This device, *fig. 2*, disconnects the electrical supply when a fault causes a current flow in the earth circuit.

The service cable is protected by a fuse which is accommodated in a box and sealed by the electricity board. This seal must never be broken by a householder, only by the board.

Fortunately, this fuse rarely blows. If it does, ring the board, but do be sure before you do so that the fault is not in your own circuits or in any plug fuse or appliance.

That is really all you need to know about the supply of electricity into your home. However, to appreciate what goes on on the consumer side of an installation, you need to know more.

So what happens after the supply has reached the consumer unit? Before we go into that, study this list of accessories and terms used in the electrical industry. It will help you to understand your circuits more fully.

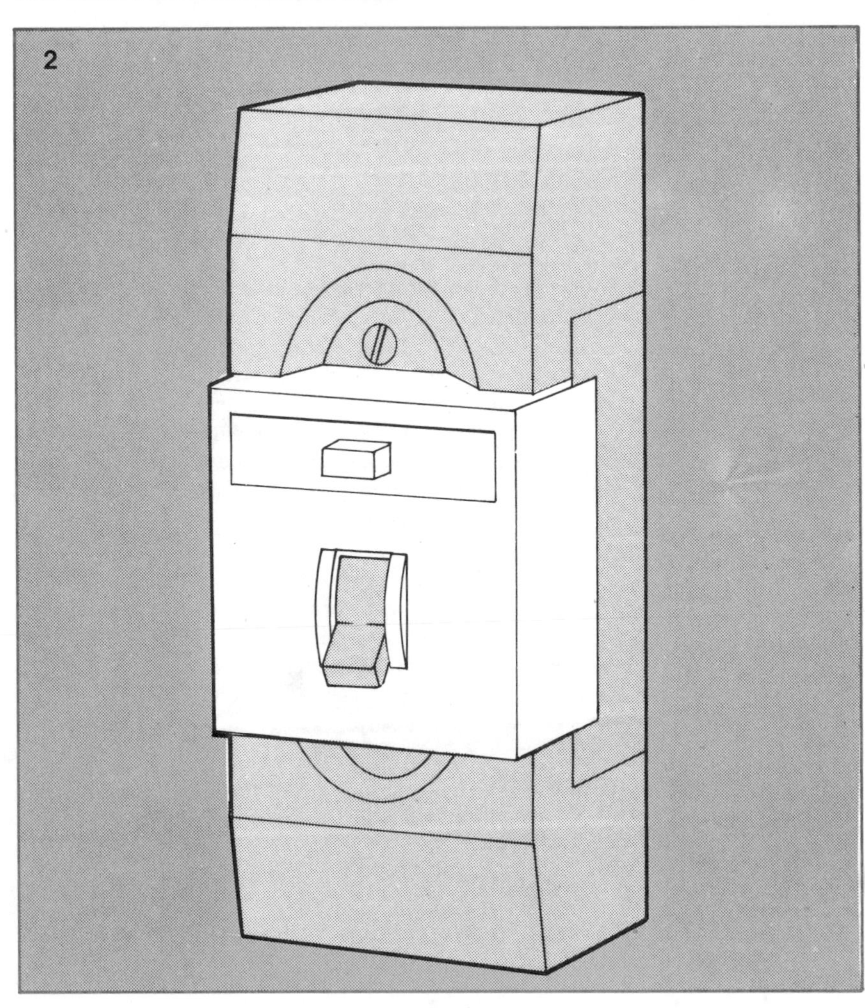

Fig. 2 Earth leakage circuit breaker

Accessories and Terms

Adaptor: Enables more than one appliance to be used from one socket outlet. Do not use too many; there is always the danger of overloading the circuit.
Alternating current: A current which changes its direction at regular intervals. Known as a.c., this is the type of current supplied to domestic premises.
Architrave box: Narrow steel box to house narrow switches. Also acts as a connection box for cables and flex wires to wall light fittings.
Backplate: Metal or plastic plate fixed behind open backed rose or switch to stop cables touching the wall or ceiling. Also *pattress.*
Blank or **blanking plate:** Covers (blanks off) a socket outlet no longer required or an outlet box used to join cables.
Block connector: Used to join cables in a wall box, or for connecting cables to flex.
Box, steel, knockout: Houses a socket outlet or fused connection unit. Is sunk in the wall to contain circuit cables.
Cartridge fuse link: Fuse element for 13 amp plugs and consumer unit fuses (alternative to fuse wire in consumer unit).
Conductors: Insulated wires in cable or flexible cords. Usually one live, one neutral; sometimes also an earth wire.
Connection unit: Outlet fitted with a fuse. Used to connect fixed appliance to ring circuit. Formerly called fused spur unit or box.
Consumer unit: Modern compact version of a fuseboard with a number of fuseways housing circuit fuses.
Core: A cable or flex conductor enclosed in insulation material. See *sheath.*
Direct current: A current which does not change direction (i.e., is continuous), as supplied by a storage battery. Also called d.c.
Earth continuity conductor: Uninsulated earth wire. The third wire of two-core and earth cable or of three-core flex. Known as e.c.c.

Flex connector: See *block connector.*
Flexible cord: Usually called flex. Lead from appliance to the source of power (i.e., socket outlet). Has one, two or three cores each containing strands of wire. See chapter on cables.
Joint box: Junction box with knockout sections permitting cables to enter and be joined to its terminals.
Knockout box: See box (steel). Also available in plastic for surface-mounted accessories.
Line: See *Live conductor.*
Live conductor: One pole of an a.c. (alternating current) two-wire electrical system. Also called line or phase. The other pole is the *neutral.* Somewhat similar to the positive of a d.c. (direct current) system.
Miniature circuit breaker: Sometimes used in consumer units instead of wire fuses. Switches off automatically if the circuit is overloaded or a fault on an appliance fails to blow the fuse in a plug or socket.
Neutral conductor: One pole of a two-wire electrical system. Resembles the negative pole of a d.c. system. See also *Live Conductor.*
Pattress: Similar to *backplate.*
Phase: Another name for live or line.
Plaster depth box: Steel box to house a plateswitch on wall.
Plateswitch: Modern term for wall switch set in slim front plate.
Rose: Accessory which screws to ceiling to house lighting cables and flex wires.
Sheath: Outer covering of cable or flexible cord.
Socket outlet: Modern version of the old 'power point'. Houses a fused plug.
Spur: Branch of a ring circuit cable which supplies fused connection units or remote socket outlets.

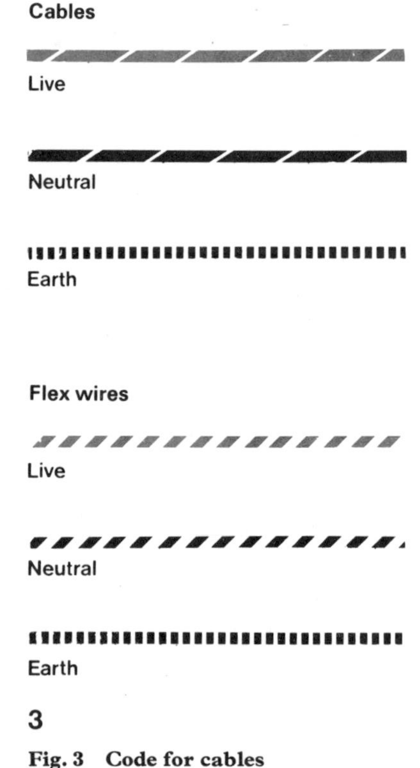

Fig. 3 Code for cables

Circuits

After the electricity supply is fed into the consumer unit, it is then distributed through the various circuits and fuses to lighting points and socket outlets (sometimes wrongly called power points).

In a typical consumer unit are a number of fuseholders housed in fuseways, *fig. 4*. The number can be anything from two to ten. Each of these holders accommodates a circuit fuse. These fuses supply and protect the various house circuits – heating, lighting, cooker, immersion heater and so on.

Each circuit is supplied by its own fuse; *no fuse may supply more than one circuit.*

Fig. 4 Connections in a typical eight-way consumer unit

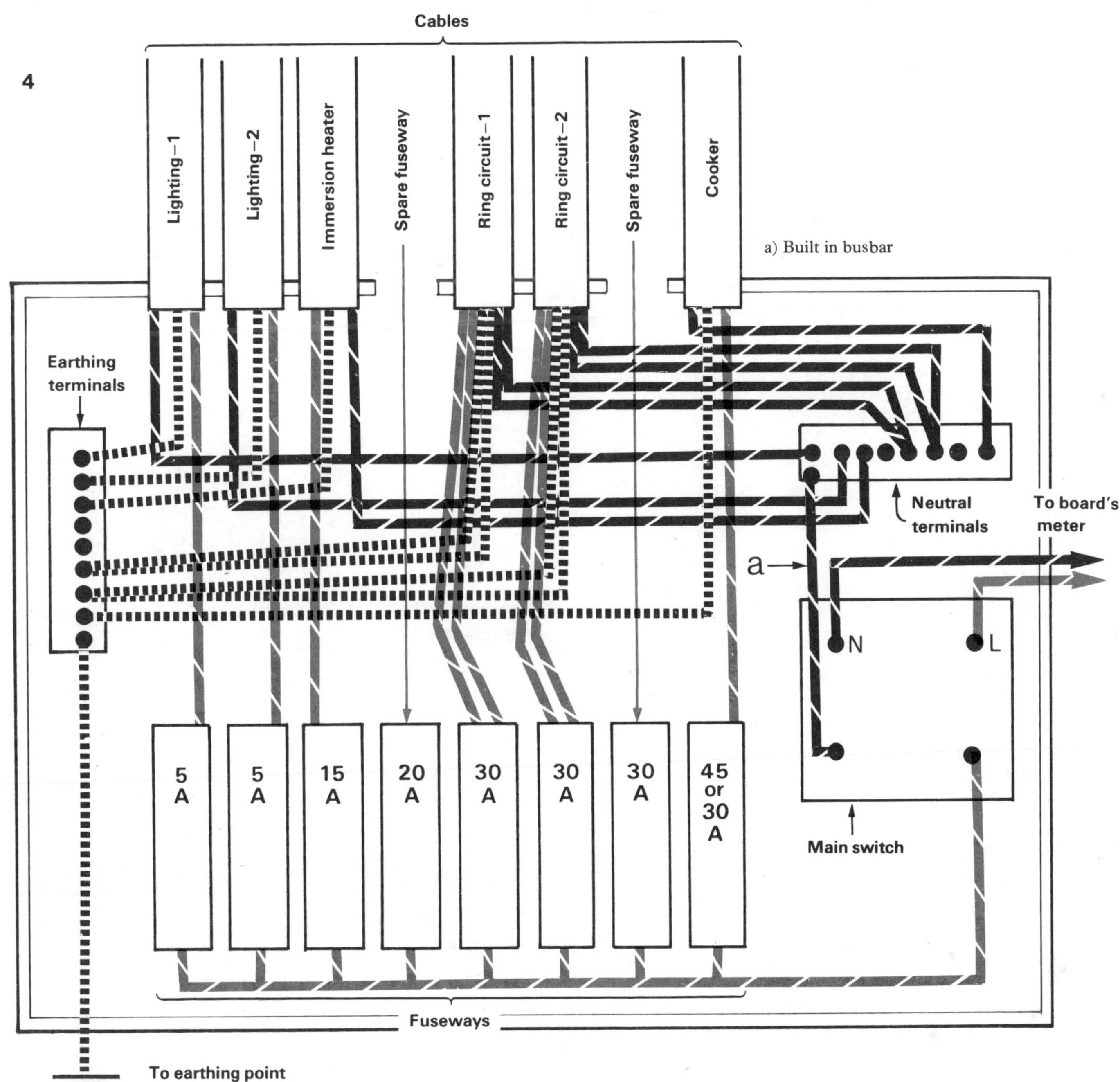

TYPES OF FUSE

There are two types of circuit fuse used in a domestic system, *fig. 6*. One type is rewirable and consists of fuse wire as the fuse element. The other is the cartridge type. In principle the cartridge fuse is similar to a 13 amp fuse used in a plug, but is bigger.

Cartridge type fuses cannot be rewired. They must be replaced if they 'blow'. No attempt to repair them should be made.

Each circuit fuse is designed to carry a maximum amount of current. This current is called the load. Each fuse has its own rating which determines the load it can carry. Generally speaking, these ratings are:

5 amp for lighting circuits; 15 amp for special circuits such as immersion heaters; 30 amp for ring or power circuits; 45 amp for large cooker and heating circuits.

FUSE COLOUR CODES

The fuses in modern consumer units are colour coded so that they can be instantly recognisable. The colours used are:

5 amp, white; 15 amp, blue; 30 amp, red; 45 amp, green.

There are also 20 amp fuses (colour coded yellow) which are used for special circuits.

In older installations, the fuse carriers may have the current ratings stamped on them in figures instead of being coded with coloured dots.

Fig. 5 Fuse box showing the amperage of the different fuses

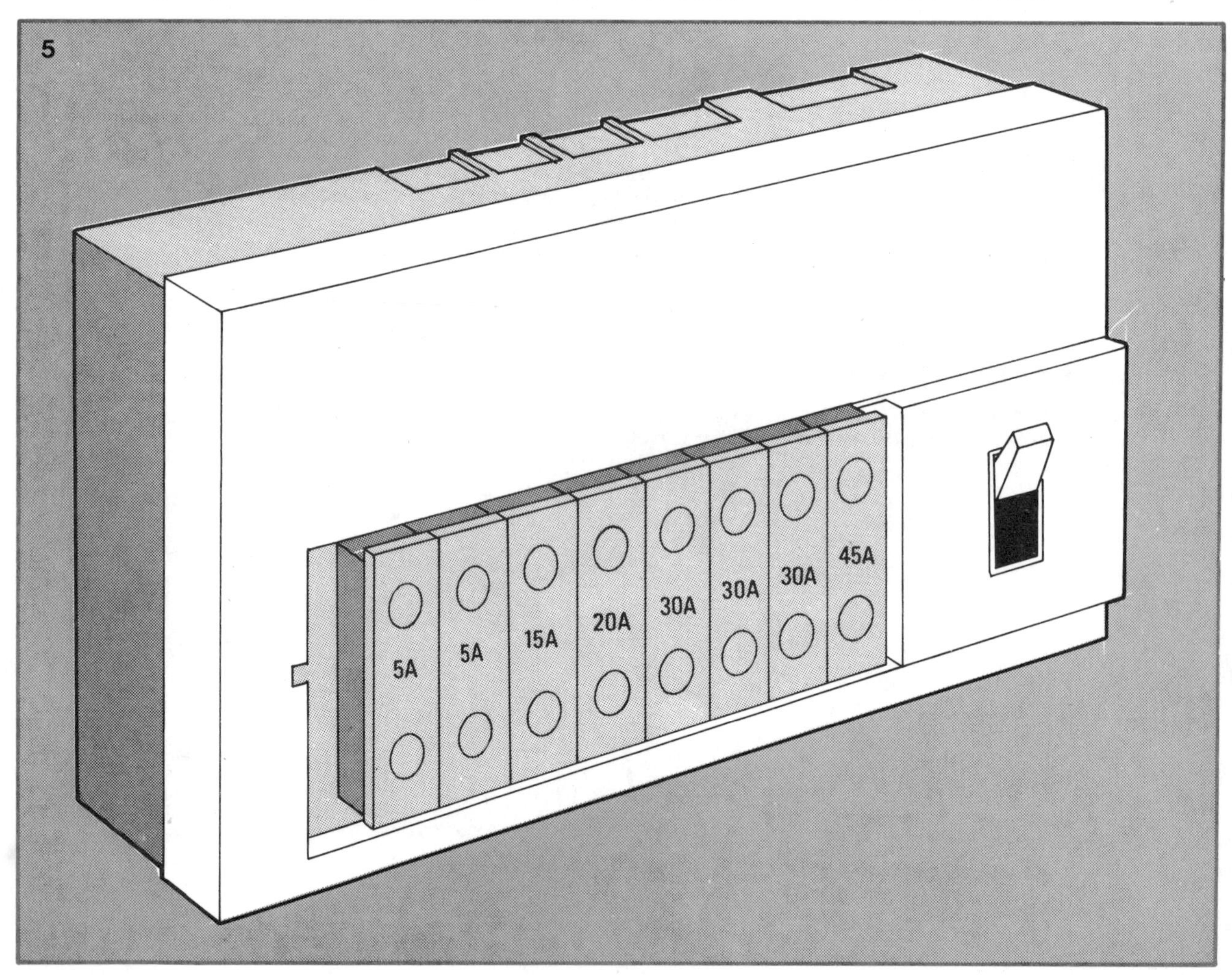

It pays to examine your consumer unit to become familiar with the fuses. If they are not easily identifiable, make a sketch of them on a postcard. Give each one a number and list its function (lighting, ring circuit, immersion heater etc.) against the appropriate number. Hang the card near the fuseways where you can find it quickly in an emergency.

If the fuses are rewirable types, keep a card of fuse wire handy. It is sold wound on small cards with a supply for each of the three main fuse ratings – 5, 15 and 30 amp.

If the fuses are cartridge types, always keep a supply of spares handy. Fuses have a habit of blowing when all the shops are shut!

Each consumer unit (I will stop referring to fuseboards from now on) has a main switch. When turned off, this switch isolates all the circuits connected to the consumer unit.

This switch should always be turned off when inspecting or mending a fuse, and when making any alteration to the wiring system. If ever you read a book or article which says, 'Turn the current off at the main switch', that is the switch you should look for.

EARTHING TERMINALS

Now for a few more details about the consumer unit itself. If you look at *fig. 4*, you will see on the left the earthing terminal strip or block. The number of individual terminals on the

Fig. 6 Circuit fuses: rewirable and cartridge types

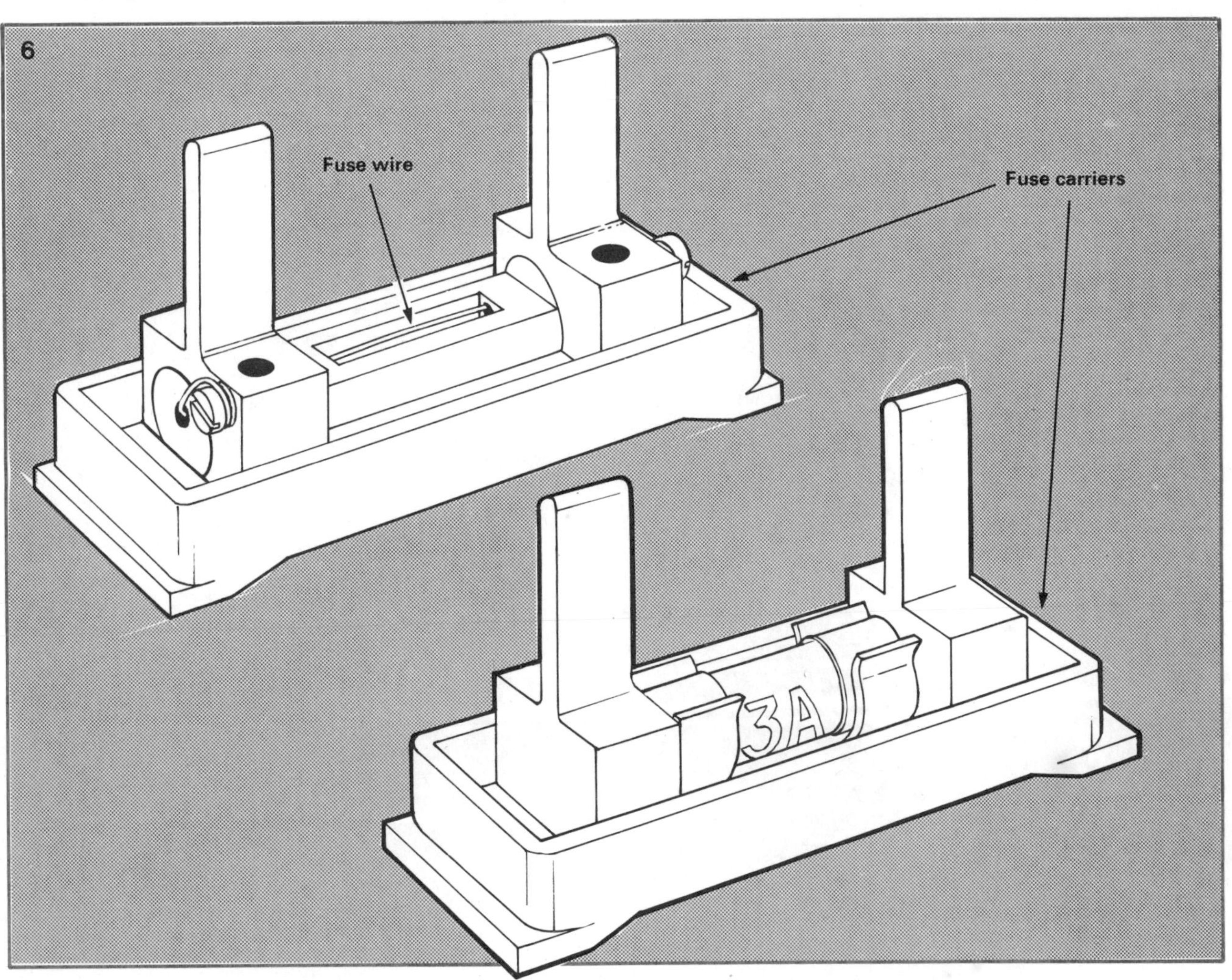

strip will, of course, vary according to the size and the make of the unit.

The terminal at the bottom of the strip is the one usually used by the electrician when the wiring of the installation is carried out. But this may not always be the case.

From this terminal runs the earth lead to whatever central earthing point the installer may have used to earth the installation.

The rest of the terminals on the strip are used to accommodate the earthing wires (conductors) of the various circuit cables running throughout the house.

NEUTRAL TERMINALS

On the right of *fig. 4* is the neutral terminal strip or block. This also has a number of individual terminals. It is to these that all the neutral conductors of the house cables are connected.

The live conductors of the cables are connected to the individual fuseways as shown in *fig. 4*. You will see that the conductors of the ring circuit cables are joined to a terminal or fuseway where another conductor has already been connected. This will be explained later.

Note that the positions of the earthing terminal strip and of the neutral terminal strip in a consumer unit can vary according to make.

CABLE LAYOUT

To make a neat wiring job, and to make it easier to locate cables, all three types of conductor should be arranged in the unit in logical order. The highest rated fuseway is now normally the one next to the main switch (farthest right in *fig. 4*) and the lowest rated is on the far left.

When wiring up a consumer unit, the earth and neutral terminals are normally used in the same order as the live fuseways are positioned. This has been done as far as it is possible in *fig. 4*.

The live terminal on the main switch is linked by a busbar to the fuseways, and the neutral terminal is similarly linked to the neutral terminal slip. These links are all part and parcel of the unit's construction and are built-in fitments.

The two leads (live and neutral) which run to the board's meter are called the meter tails and are connected to the meter by the electricity board when they have approved the installation. This must only be done by the board.

If you ever have to install a consumer unit, you will be required to supply and connect these two tails to the unit's main terminals, so that they are ready for the board to connect to the meter.

The tails should normally be about one metre in length, but if for any reason the consumer unit is not positioned near the meter (as is the normal procedure) the tails will, of course, need to be longer.

Cables and Flexes

Fig. 7 **Examples of cables and flexible cords**
a) Mineral insulated copper clad (m.i.c.c.)
b) PVC armoured
c) PVC sheathed and insulated (house wiring cable)
d) As above, but seven-strand if 4.0 mm² or above
e) PVC sheathed and insulated twin flat
f) PVC parallel twin (figure of eight)
g) Cotton braided circular
h) PVC twisted twin

Although there are many types of cables and flexible cords (usually called flex), for general wiring purposes the householder needs to be concerned with only a few.

First, however, it is important to be quite clear on the difference between cable and flex. A cable is used for fixed wiring; flex is used to connect a portable appliance to a plug; a lampholder to a ceiling rose; or a fixed appliance to its fused outlet.

Cable should never be used instead of flex, and flex must not be used as a substitute for cable. This is a very important rule.

CORE COLOURS

Another point – an important one – to remember is the colour of the cores (insulated conductors). Although the colours of three-core flexible cords were changed some time ago, house wiring *cable* core colours were *not* changed.

The colours are:

CABLES: Live is red; neutral is black; earth is green.

FLEXIBLE CORDS: Live is now brown; neutral is blue; earth is green with yellow stripes.

The *old* colours were: Live (red); neutral (black); earth (green). On older appliances you will probably find the flex wires are in the old colours.

The *sizes* of both cables and flex have been changed to metric measurements. *Table A* gives some cables sizes and *Table B* gives the sizes of some flexible cords. *Figure 7* shows some examples of cables and flexible cords.

As a general rule, it is necessary for a householder to remember only two or three sizes of cable. These are 1·0 mm.² (or 1·5 mm.²) for lighting circuits; and 2·5 mm.² for ring circuits, 20 amp storage heater circuits and immersion heaters. For cooker circuits 6 mm.² or 10 mm.² cables are used.

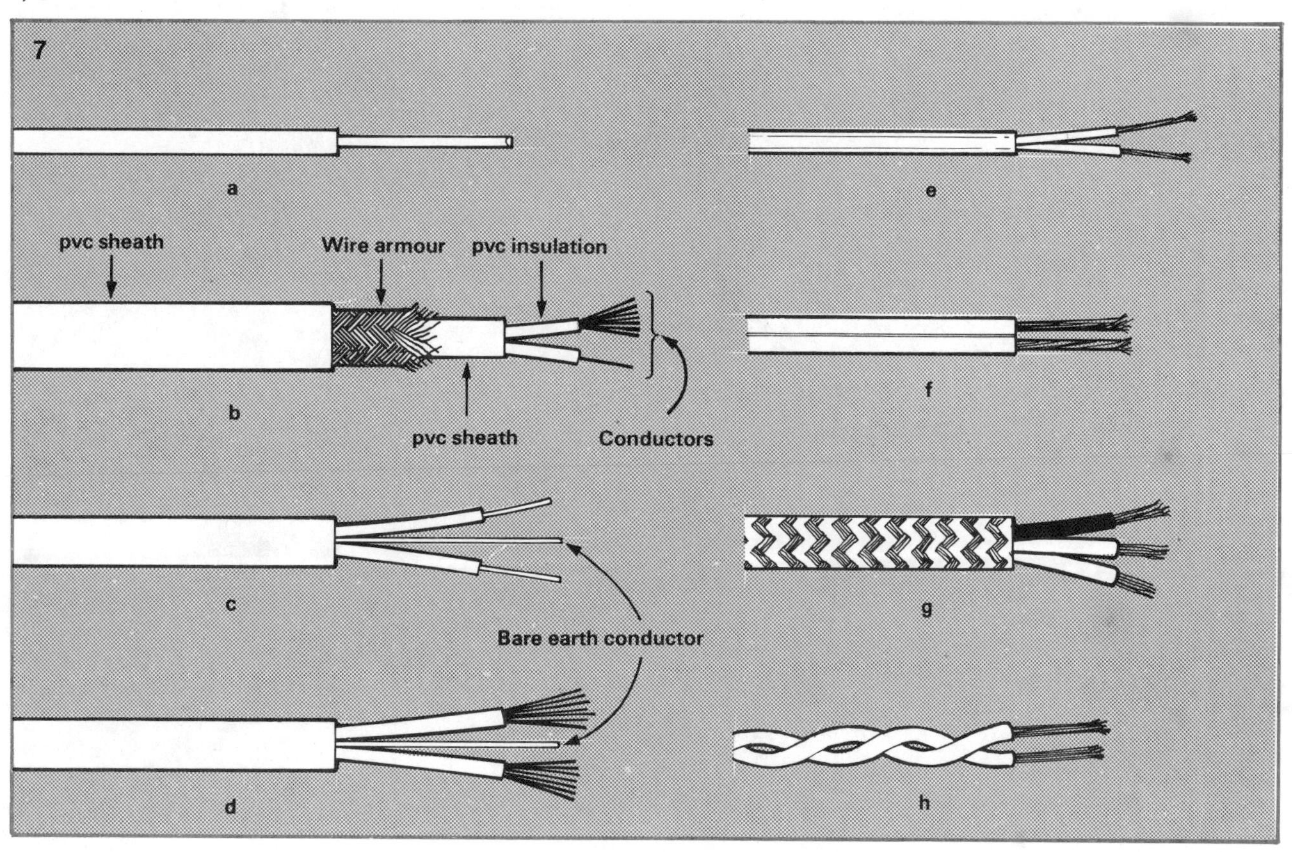

For interior house wiring, pvc sheathed cable is normally used. It is available in single core (one conductor); twin core; twin core and earth; and three core.

For outside wiring, there are two types of cable especially suitable. One is m.i.c.c. (mineral insulated copper clad). The other is armoured pvc insulated twin-core cable. See separate chapter on outside wiring.

TABLE A
Cables

Cable size	*Max. current rating (amps)*	*Circuit fuse*
1·0 mm.^{2}	12	5 amps
1·5 mm.^{2}	15	15 amps
2·5 mm.^{2}	21	30 amps

TABLE B
Flexible cords

Cord size	*No. of wires*	*Current rating*
0·5 mm.^{2}	16	3.0 amps
0·75 mm.^{2}	24	6·0 amps
1·0 mm.^{2}	32	10·0 amps
1·5 mm.^{2}	30	15·0 amps

The main uses for the various sizes of flexible cords include:

Lighting, 0·5 mm.^{2}

Lighting and small appliances, 0·75 mm.^{2}

Appliances rated up to 2 kW, 1·0 mm.^{2}

Appliances rated up to 3 kW, 1·5 mm.^{2}

It will be seen from *Table B* that flexible cords have a number of fine wires (strands) which combine to make up the conductors. The reason for this is to make the flex as flexible as possible.

Cables used to have a number of strands, too, but most of them are now single strand. Many houses, however, are still wired in the old multi-strand cables.

For circuits such as cooker circuits which carry heavy current, seven-strand cables are the ones which are used.

WIRING HINTS

If a lot of wiring is planned, it will pay you to buy a proper cable stripper. These are available at modest prices. Take care, if you use a knife to remove the sheath or the insulation material from a cable or flexible cord, not to cut the conductor wires.

The best way to join cables is to use a joint box (sometimes called a junction box) where the ends of the conductors can be safely anchored under its terminals. A joint box suitable for lighting circuit cables is shown in *fig. 72*.

Fig. 8 Typical cable stripping tools

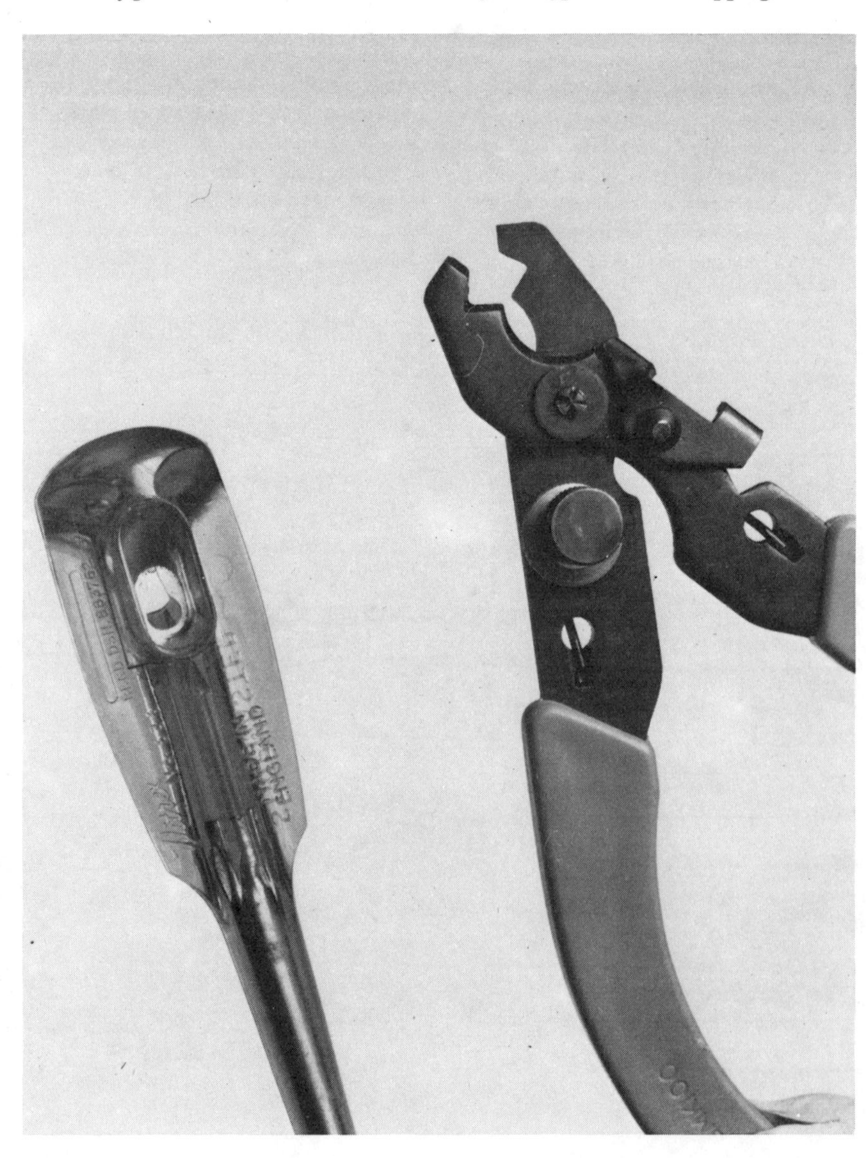

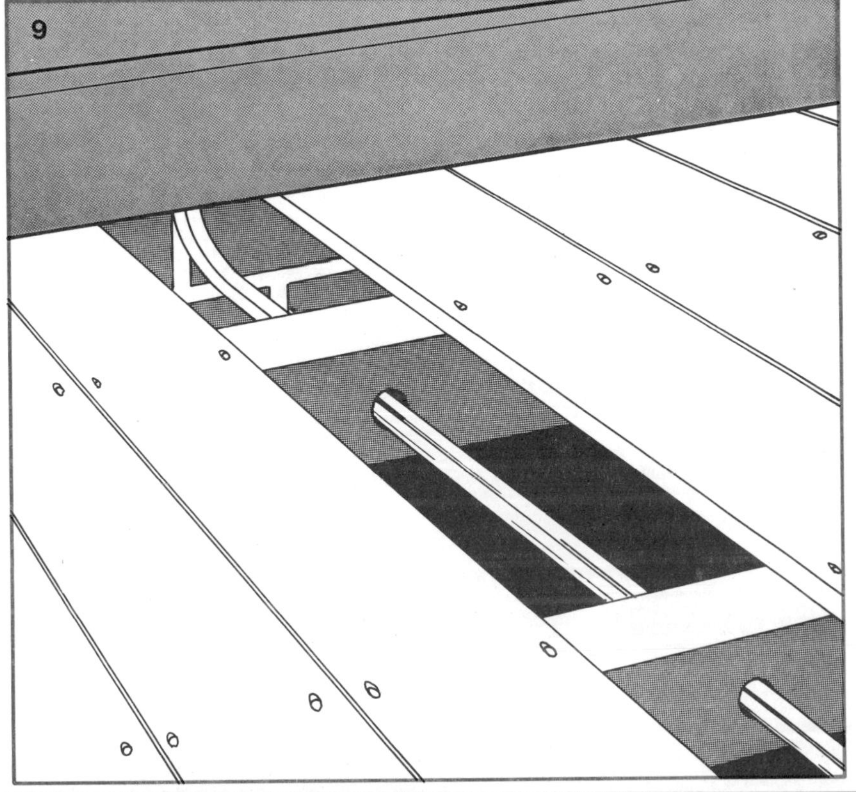

PVC (polyvinyl chloride) sheathed cable can be buried under wall plaster. There is no need for further protection.

Where cables cross timber joists under floorboards they should pass through holes drilled not less than 2 in. below the tops of the joists.

Cables should not be laid in grooves cut in the tops of the joists. The danger here is that nails driven through the floorboards could penetrate the cables.

Do not use insulating or self-adhesive tape to join flexible cords or cables. Proper flex connectors and cable connectors (*fig. 10*) are available which ensure a sound and safe connection.

Fig. 9 Cables should pass through holes in the joists

Fig. 10 Flex connectors

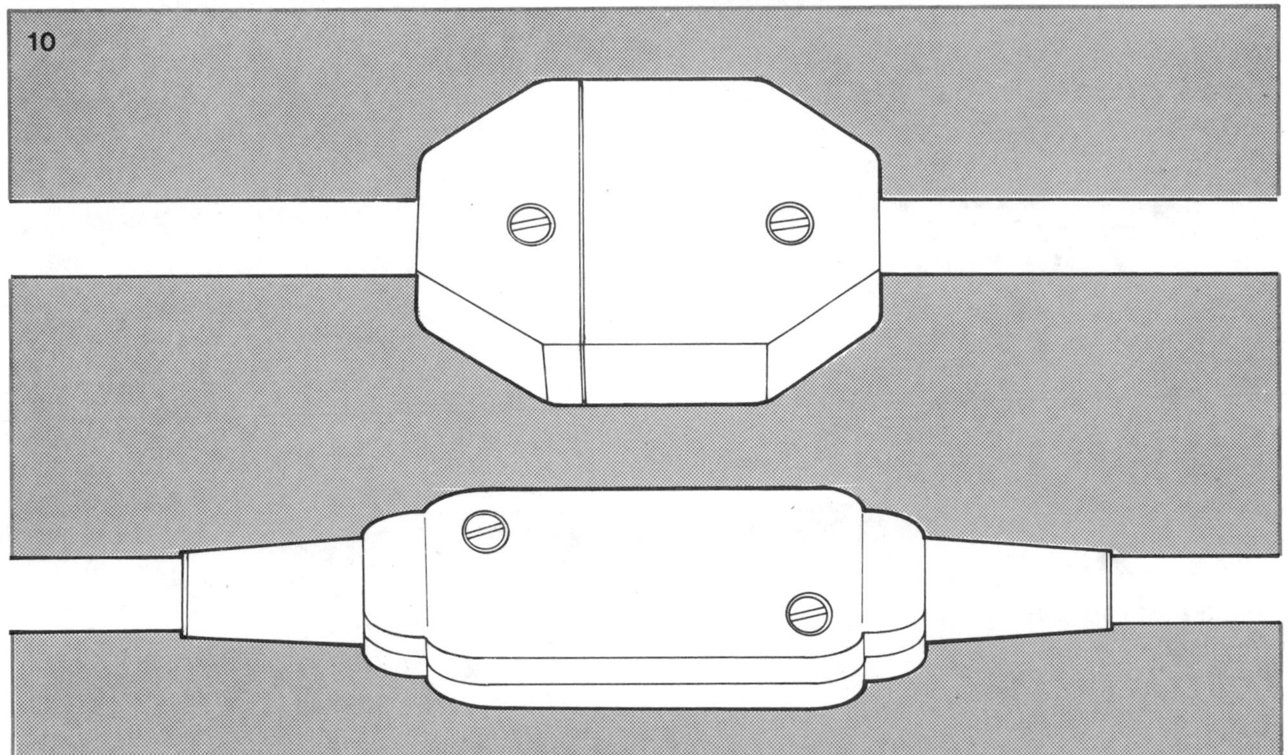

Mending Fuses

A common fault which occurs in a house wiring system is the failure of an electric light. Being suddenly left in pitch darkness can be unnerving, even dangerous, so always keep a candle or a torch where it can be found quickly.

When a light fails, the normal reaction is to replace the lamp. But in these days of power cuts, especially, it is a good idea to check first if any lights are showing from neighbouring houses.

If they are ON, then try your other lights *on the same floor*. The reason for specifying the same floor is that the house probably has two or even three lighting circuits – one for each floor is the common practice.

CHECK THE MAIN FUSE

If the lights on the same floor do not work, the lighting circuit fuse in the consumer unit may have blown. Turn off the main switch in the consumer unit and check the appropriate lighting fuse by pulling it from its holder to inspect it.

In rewirable types, the fuse wire element will be broken if the fuse has blown. Here's how to mend it.

Release the two screws and remove the broken fuse wire – *all of it*. From your card of fuse wire, thread the 5 amp wire through the hole in the fuse holder at one end and tighten the wire under the screw (clockwise). Trim any surplus wire around the screw.

Cut the fuse wire to the length you require and fix this end under the other screw. Do not stretch the wire or allow it to become too slack. *Figure 6*. page 9 shows this type of fuse.

If the fuses are cartridge types, they should simply be renewed if suspect. They can be tested as shown in *fig. 11*.

After repairing or renewing the fuse, replace it in its holder and turn on the main switch. If the fuse blows again immediately the light is switched on, there is a fault somewhere which needs investigation. This could be a broken or disconnected flex wire at the bulb holder.

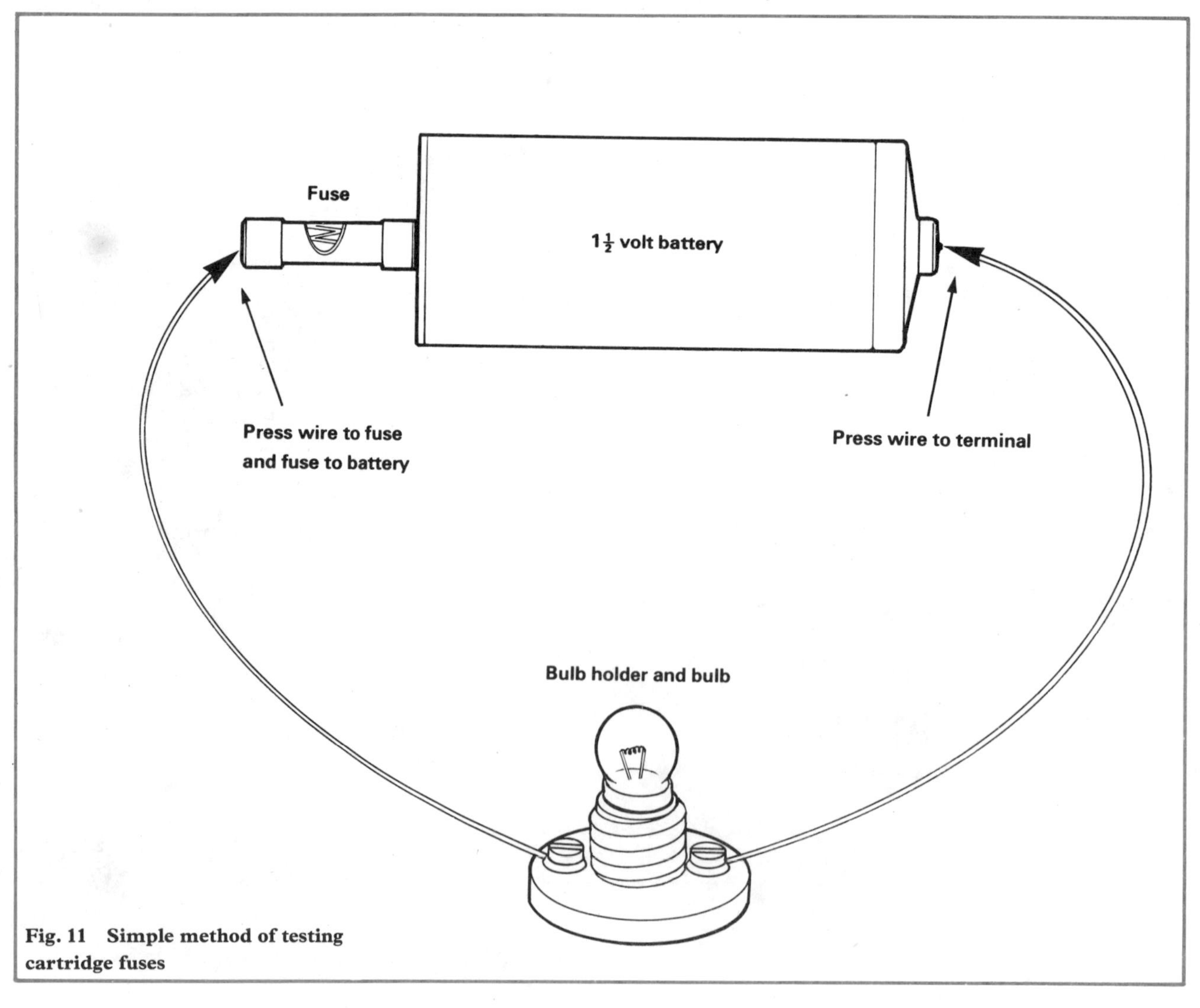

Fig. 11 Simple method of testing cartridge fuses

If the lighting fuse has not blown, and the bulb is in order but still does not light up, check to see if the ring circuit is working. You can do this by switching on an appliance which you know is in order. If that fails to work also, then the electricity board's main fuse may have blown. Telephone the board only when you have eliminated all possible causes of the fault on your side of the installation.

To reduce the risk of being left in the dark when a lighting circuit fuse blows, it is a good idea to have a standard or other type of lamp plugged into a socket outlet on the ring circuit in as many rooms as you may find to be practicable.

FUSES TO USE

When acquiring an appliance, make sure you know its wattage and the correct fuse to use in the plug. The following is a rough guide to the rating of fuse to use for appliances in common use. If in doubt, always ask in the shop or follow the manufacturer's instructions.

Appliance	*Approx. wattage range*	*Fuse needed*	
		Flat pin plug	*Round pin plug*
Blanket	150 or less	3 amp	2 amp
Coffee maker	250 to 750	3 amp	5 amp
		13 amp (750 W)	5 amp
Cooker	to 3 kW	13 amp	15 amp
	(fuses of larger cookers wired in circuit)		
Drill	150 to 450	3 amp	5 amp
Food mixer	to 300	3 amp	5 amp
Heater (radiant)	to 3 kW	13 amp	15 amp
,, (convector)	to 3 kW	13 amp	15 amp
,, (fan)	2 to 3 kW	13 amp	15 amp
Hair drier	to 720	3 amp	5 amp
,, ,,	over 720	13 amp	15 amp
Iron	750 to 1,200	13 amp	5 amp
Kettle	750 to 1 kW	13 amp	5 amp
,,	1 kW to 3 kW	13 amp	15 amp
*Refrigerator	120	13 amp	5 amp
Spin drier	200	3 amp	5 amp
*Vacuum cleaner	150	3 amp	5 amp
,, ,,	to 750	13 amp	5 amp
Washing machine (motor 300 watts; heater 2½ kW)		13 amp	15 amp

*Refrigerators and some vacuum cleaners draw a large amount of current when starting up and so require a plug fuse of a slightly higher rating than is usual for their wattages.

Jobs Around the House

There are quite a few simple wiring jobs which need to be done regularly in any household. Several of them are outlined here.

Always make sure before you start that you have all the tools and materials you are likely to need. Don't follow a make-do-and-mend policy. It simply does not pay where electricity is concerned. Take your time over wiring connections and don't be tempted to experiment.

Always use the correct size cable or flex for the job, not just any odd bit of wire that may be lying around. Remember, electricity can be lethal, so **TAKE CARE.**

NEW FLEX FOR AN IRON
Friction often causes the flexible cord of an electric iron to become worn and therefore potentially dangerous. If the sheath of the flex is worn near either end, it may not be necessary to replace the whole length. If the sheath is worn nearer the middle, however, the complete flex should be replaced without delay.

If it is not renewed, before long the insulation of the conductors will also become worn and the wires will be exposed.

Fig. 12 Tools for the home electrician

Fig. 13 How an electric iron is wired

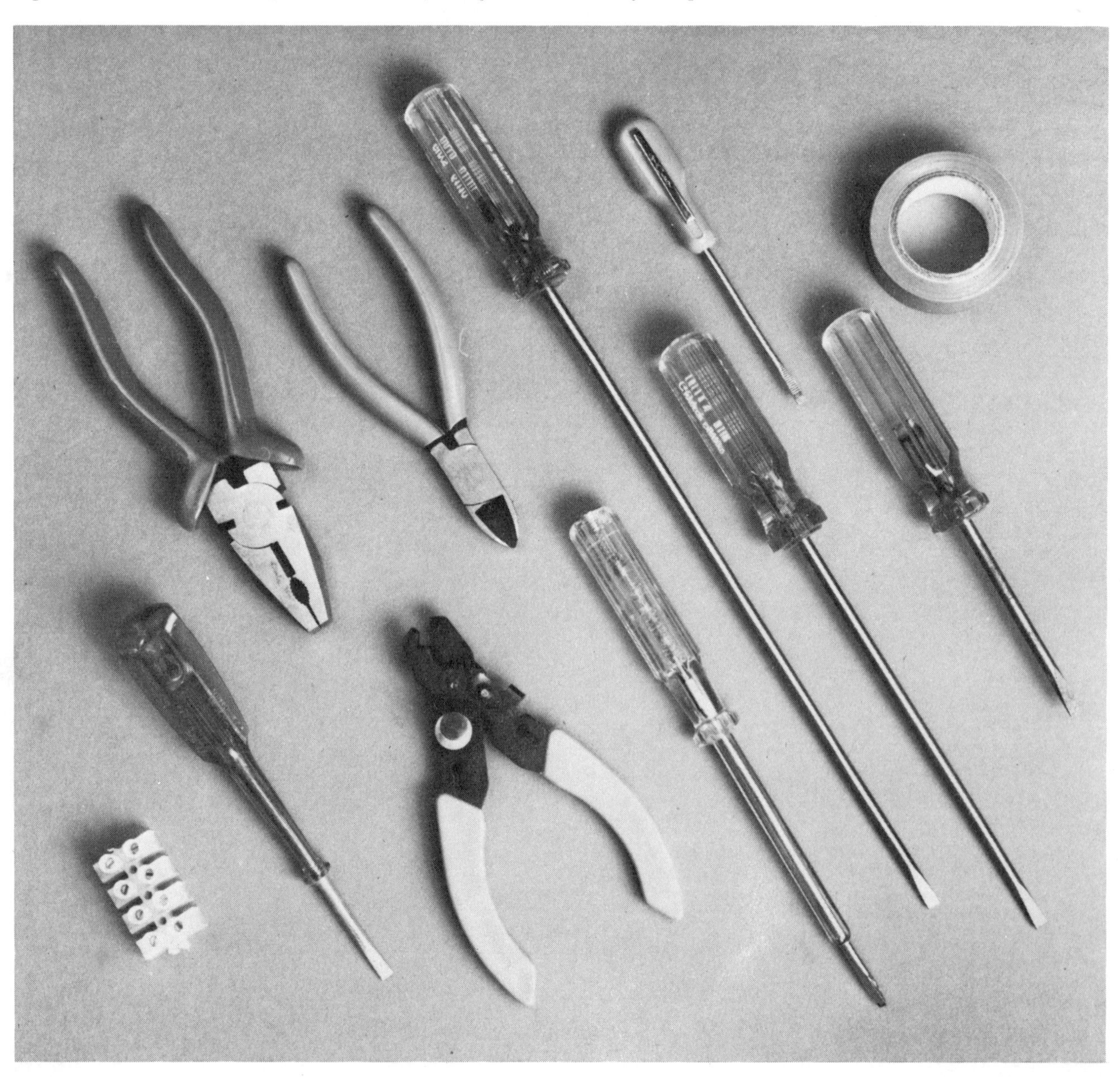

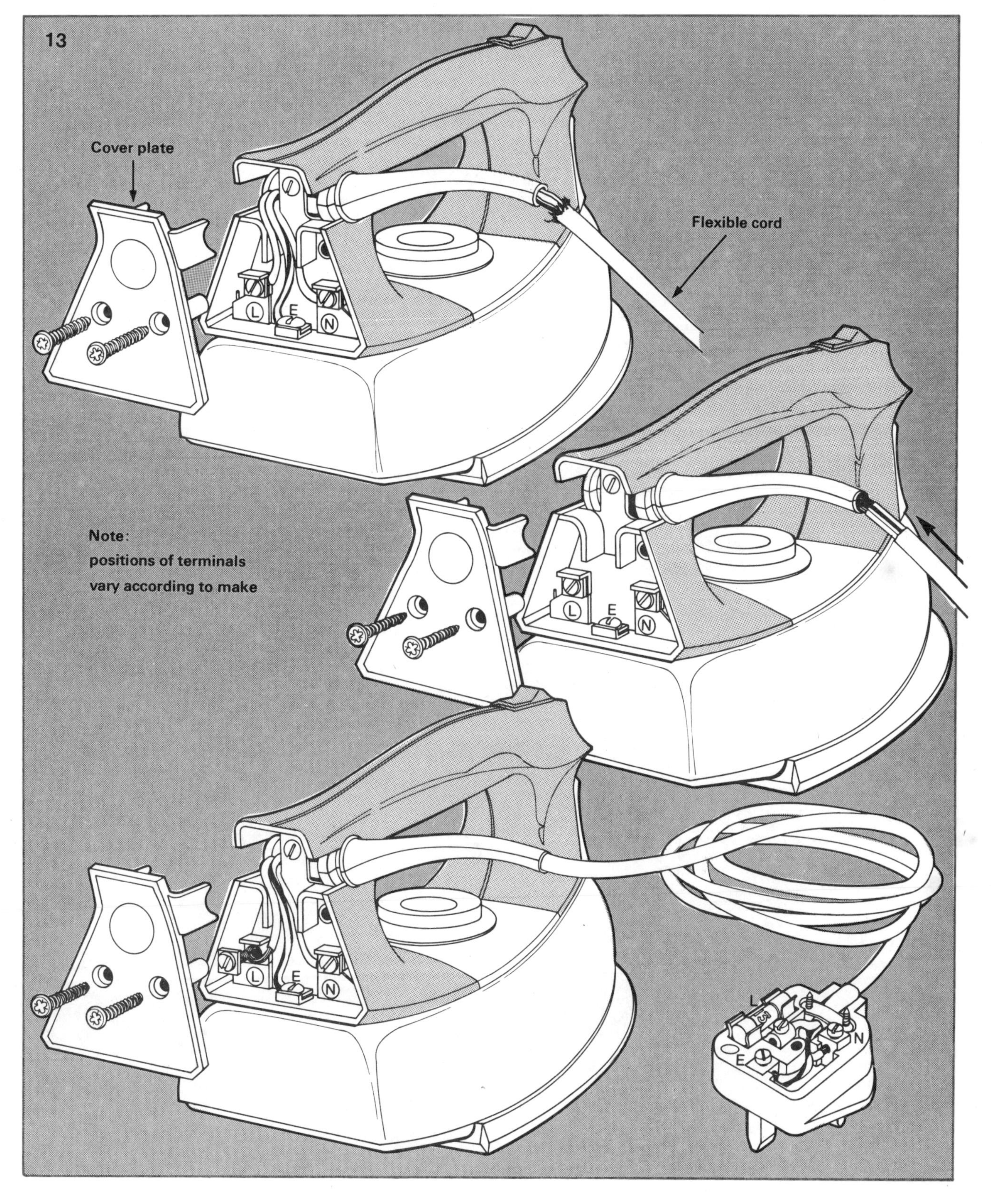
13
Cover plate
Flexible cord
Note:
positions of terminals
vary according to make
L
E
N

Replacing the flexible cord is not difficult, but it is most important to make a sketch of the wiring arrangements of the iron before disconnecting the conductors from their terminals.

In most cases, these terminals will be located under a cover at the back of the iron (*fig. 13*). The arrangement, however, will depend on the make. So first take off the cover and make your sketch.

Then release the three conductors from their terminals and pull the old flex through the grommet. Make sure no loose strands of the flex are left on or around the terminals as these could cause a short circuit.

Insert one end of the new flex (you will need about 9 ft. of unkinkable flex) into the grommet. Strip off a suitable length of sheath. Then prepare the ends of the three conductors. You can use the old flex as a guide to determine how much wire to leave exposed.

The blue core of the flex goes to the neutral (N) terminal; the brown core to the live (L) terminal; and the green/yellow core to the earth (E) terminal.

(On an old iron, red, black and green were the colours used for live, neutral and earth respectively, but as stated earlier, these colours were changed some time ago to fall in line with world standards.)

Make a final check that the connections have been properly made. The ends of the conductors must be firmly anchored under the terminals.

The sheath should now be anchored in the same way as it was originally. Replace the terminal cover. Then connect a fused plug fitted with a 13 amp fuse to the other end of the flex. If you have no ring circuit and use round pin plugs, fit a 5 amp type.

WIRING UP PLUGS

Fitting a plug to the end of a flexible cord is not a difficult job, but care must be taken to ensure that the wires are fitted to the correct terminals. This may sound obvious, but it is very easy to make a mistake.

To wire up a three-pin fused plug, first take off the cover by removing the centre screw located at the back of the plug. Release the cord grip (if there is one) by removing the other two screws on the back of the plug.

Fig. 14 Connections for a 13 amp plug with pillar type terminals

Fig. 15 13 amp plug with clamp-type terminals

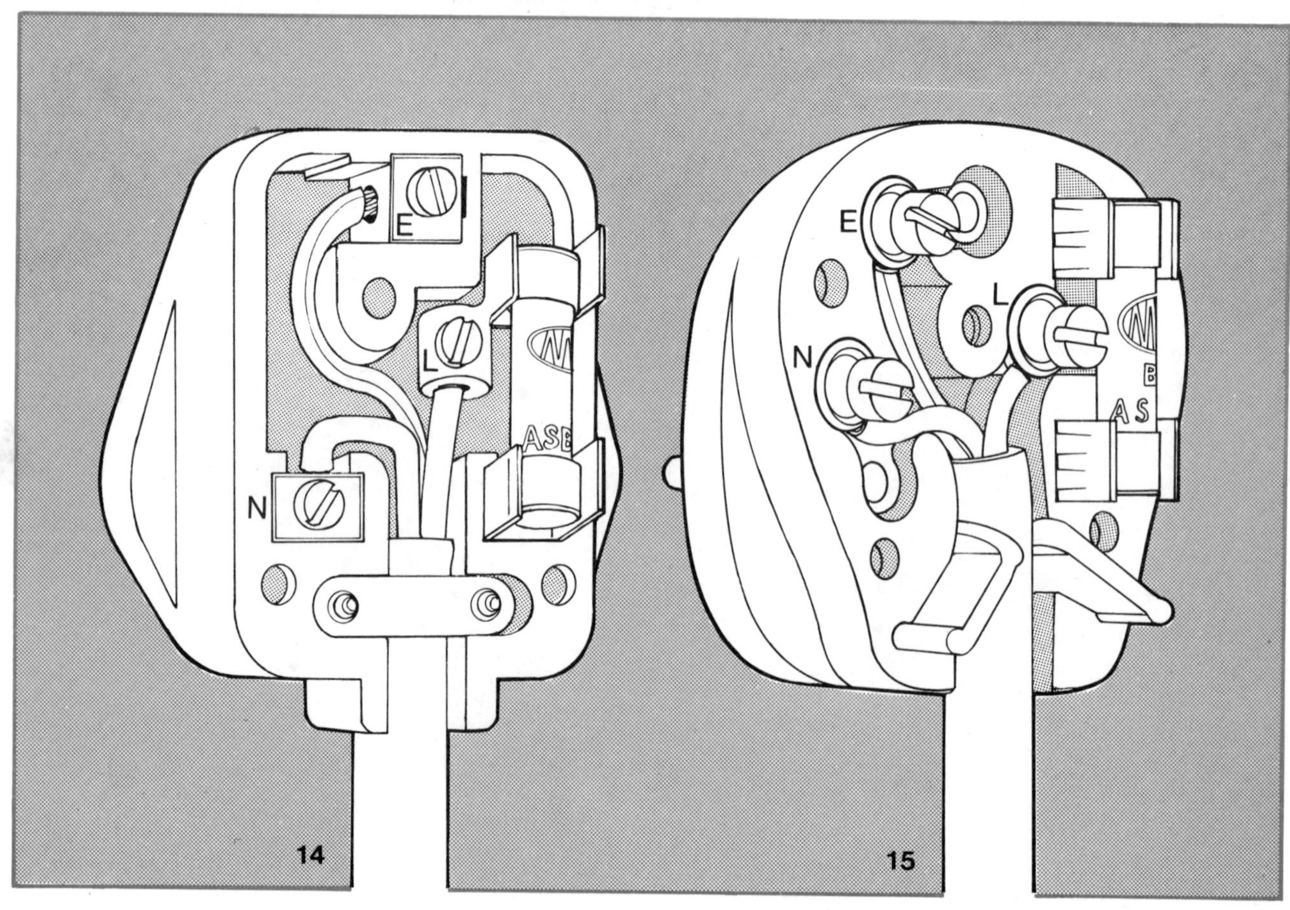

Now take out the cartridge fuse by levering it gently from its sockets. On plugs with pillar type terminals this probably will not be necessary.

Remove about $1\frac{1}{2}$ in. of the outer sheath from the flexible cord and $\frac{1}{2}$ in. of insulation material from the ends of the three conductors. Connect the conductors as in *figs 14 and 15.*

It is important that all the strands of each conductor are twisted together as a stray strand could foul the wrong terminal.

If the terminals are the clamp type, wind the bare ends of the conductors in a clockwise direction under the terminal washers. Tighten the screws.

Pillar type terminals are those shown in *fig. 15.* The clamp types are illustrated in *fig. 14.*

To get a firm fixing in pillar type terminals, double back the ends of the conductors on themselves, insert in the terminal holes and tighten the small screws on top of the terminals.

Now anchor the flex by replacing the cord grip and tightening its screws. Before replacing the cartridge fuse, check that it is of the correct value for the appliance.

For appliances which are rated up to 720 watts, choose a 3 amp fuse. If the appliance is rated at between 720 and 3,000 watts (3 kilowatts), insert a 13 amp fuse. A table of fuses for various appliances is given on page 15.

Arrange the conductors neatly in the plug (slots are usually provided), replace the cover and tighten the centre screw on the back.

Some modern plugs have an alternative method to the cord grip for anchoring the flex.

READING THE METER

Every householder should be able to read his or her electric meter. For one thing it helps you to keep a close watch on the amount of electricity you are consuming. For another, you may be asked by your board to provide a reading yourself if you should be out when the man calls to read the meter.

A typical meter is shown in *fig. 16.* When taking a reading, ignore the small dial marked 1/10kWh. The other dials are then read from left to right, although there seems to be two schools of thought about this.

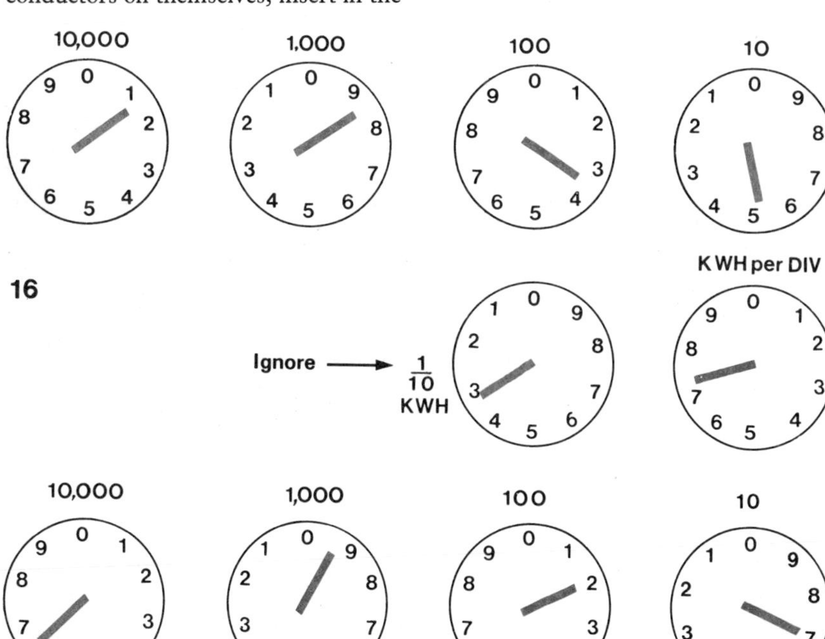

Fig. 16 Ignore the dial marked 1/10 kWh when reading a meter

Fig. 16a Note carefully the positions of the pointers

Some say read from right to left, putting the second figure down on the left of the first and so on. Personally, I feel that there is always the possibility of obtaining a reading in reverse by using this method! However, you have a choice.

Using the left to right method, note that the digit to read is the one which the dial pointer has just *passed*, not necessarily the digit which is nearest the pointer.

Deduct the previous reading shown on your last bill for electricity from the reading you have taken. The answer will be the number of units consumed since the meter was last read. This, of course, can be done at any time if you want to check consumption over a given period.

Some meters have no 10,000 dial, but the method of reading is the same. The reading of the meter in *fig. 16* is 18,357.

ERRORS TO AVOID

It is possible to take an incorrect reading of a meter by failing to note carefully the positions of the pointers. To illustrate my point, look at *fig. 16a*.

At a quick glance it would appear that the reading is 69,273. On closer examination, however, we see that although the 10,000 dial pointer has passed figure 6, the 1,000 pointer is between 9 and 0 and has not, therefore, completed its revolution.

The correct reading is 59,273, which makes a considerable difference!

FITTING A KETTLE ELEMENT

To fix a new element in an electric kettle, first unscrew the shroud ring, *fig. 17*. This may be tight, so you may need a wrench to shift it. Avoid damaging the chrome of the kettle by protecting it with a thick cloth.

Next, remove the washer immediately behind the ring. Push the flange towards the kettle and lift out the element through the top. Remove all traces of the old washer and scale from both sides of the element hole.

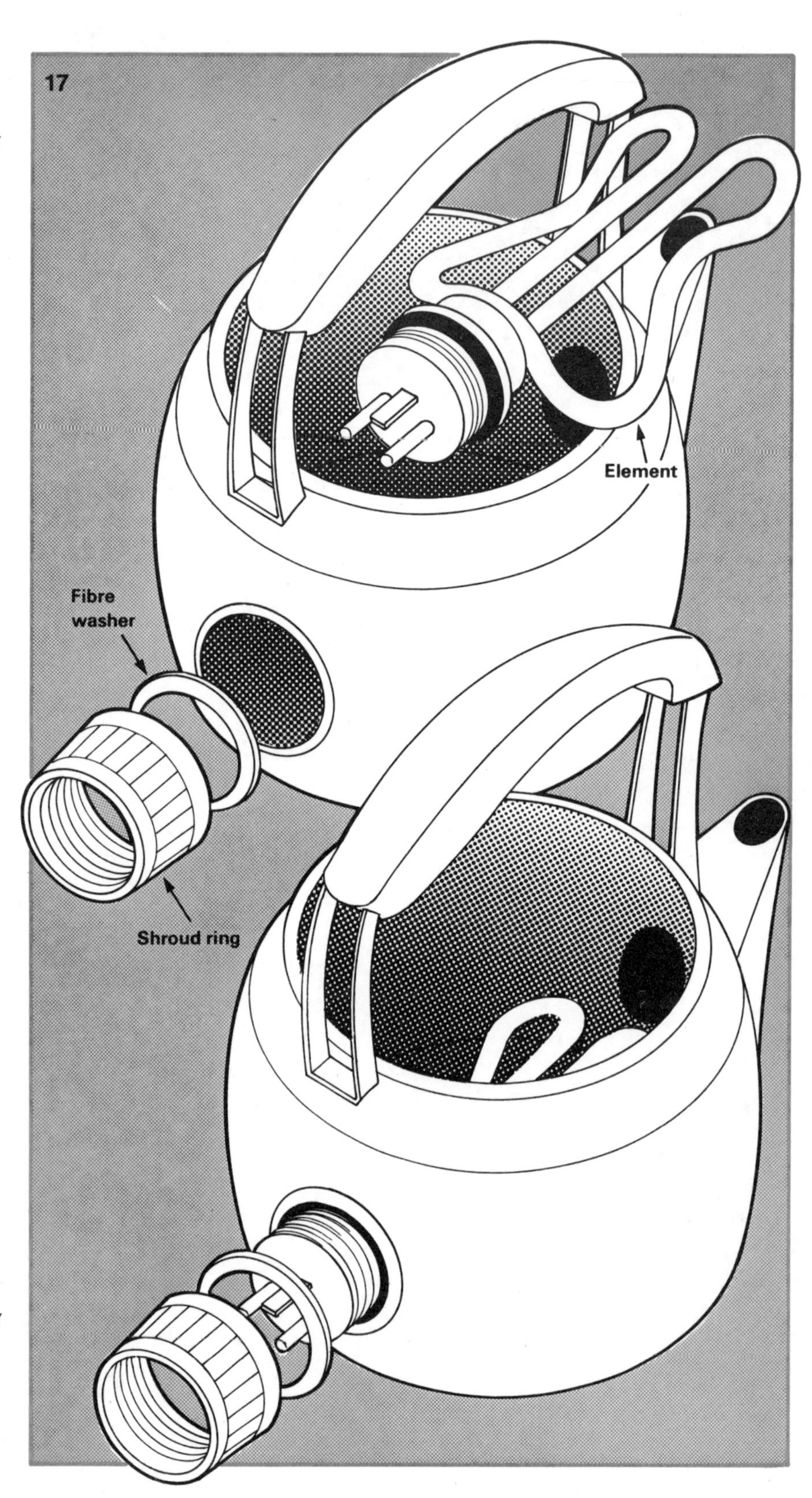

Fig. 17 Remove the shroud ring and washers to renew a kettle element

There are several types of element, so unless you know exactly the one you want, take the old one to the shop as a guide. With the new element buy a fibre washer and a rubber washer.

Put the rubber washer on the flange of the new element. This washer is the one which goes inside the kettle. Pass the element through the hole into the kettle the right way up. Pull the flange through the hole and position the element.

Watch for leaks Now fit the fibre washer on the flange and screw the shroud ring on tightly so no water can leak out.

If the kettle leaks slightly when filled, try tightening the shroud still more. Be careful if you use a wrench not to overdo it and damage the rubber washer inside.

Make sure the water covers the element before you push the connector in and switch on.

Sometimes when a kettle boils dry it damages the connector and the flex. If this happens, the best thing to do is to buy a kit which contains an element, washers and a connector with the flex already wired into it.

REPLACING LAMPHOLDERS

Sometimes bayonet type lampholders become damaged and need replacing. Or the contact plungers, *fig. 18*, can lose their tension. The plungers are the two pins whose ends make contact with the lamp.

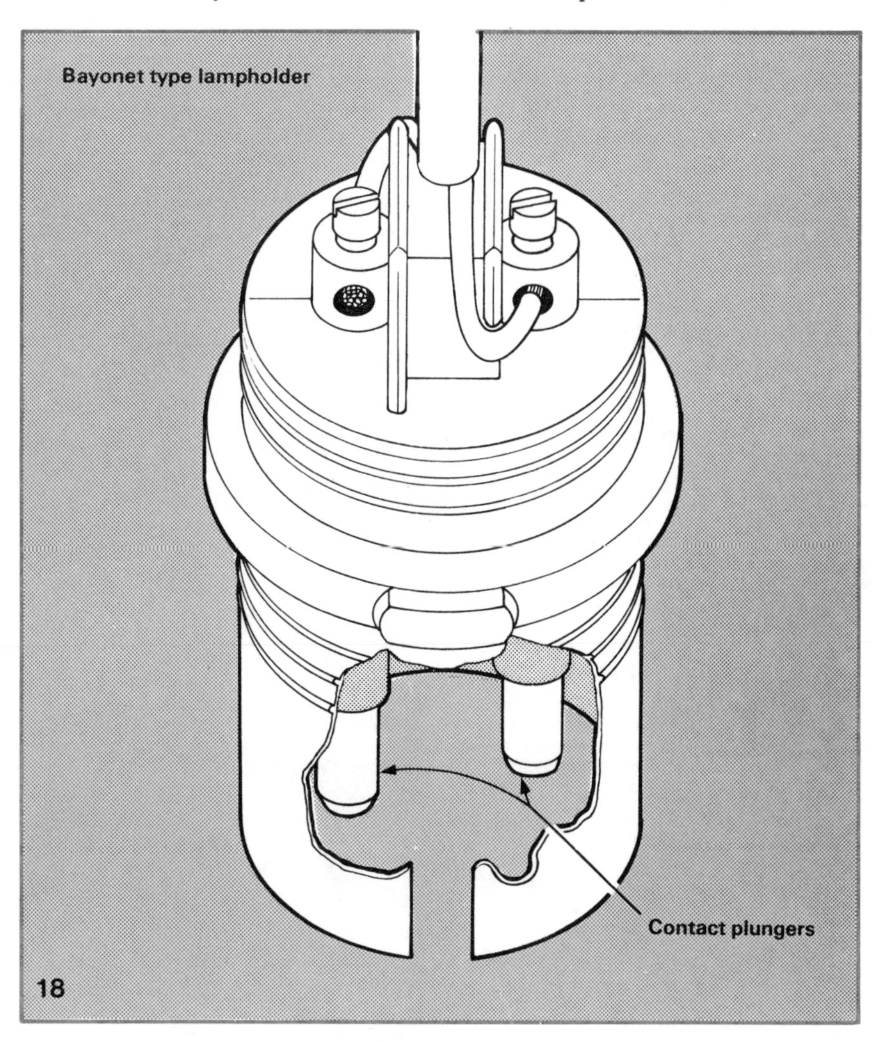

Fig. 18 Contact plungers in a lamp holder can lose their tension

Before attempting to replace the holder, turn the current off at the main switch.

Take out the lamp, remove the shade and unscrew the shield or cover, *fig. 19*. Disconnect the flex from the two terminals. If the shade or the fitting is metal, there should be an earth wire fitted. This, too, will have to be disconnected.

Now take off the cover from the new lampholder and pass the flex through it. Connect the flex to the terminals and make sure the screws are tightened firmly.

Arrange the flexible conductors neatly around the pillar of the holder. Screw the cover on to the holder and replace the shade.

The shade is normally held in place by a skirt or a plastic ring. Take care not to get the ring or skirt on a cross thread and do not overtighten. If you do, the ring or skirt will be very difficult to remove next time. Put the lamp in, restore the current and switch on to test.

WORN FLEX

If the flex is worn, this too should be renewed. But if the damage is confined to the end of the flex, it may be sufficient simply to cut an inch or so from the worn end.

With ceiling lights this obviously will depend on the total length of the flex and how close the fitting is to the ceiling.

If, however, the whole length needs to be renewed, it will be necessary to unscrew the cover of the ceiling rose. This may be a difficult thing to do and it may be necessary to break the rose in order to get it off. This will mean fitting a new rose and to do this you will have to disturb the cables themselves.

So before you break the rose, do make sure you are experienced enough to carry out the work involved.

When the cover of the rose is removed you will see the cables leading through a hole in the ceiling. Don't disturb them if you are merely connecting a new length of flex. Simply remove the old flex from its two terminals and connect the new flex, making sure you tighten the terminals firmly. *Figure 21* shows one arrangement of cables and flex at a ceiling rose. Your rose may not necessarily look like this, so make sure you connect the flex to the proper terminals.

If you have to replace the old rose, make a careful note of the cable and flex connections before you disconnect any wires. After loosening the terminals and releasing the wires, unscrew the fixing screws holding the base of the rose to the ceiling and withdraw it.

Thread the cable ends through the base of the new rose and connect their ends to the proper terminals. Note that the red cable or cables (there will probably be more than one in a terminal) go to the live terminal and the black cable or cables to the neutral terminal.

In some installations the apparent confusion of wires at a ceiling rose has to be seen to be believed, so I repeat, make sure you know which wires go where before you disconnect any of them.

EARTHING

To conform with the wiring regulations, an earth wire (a single length of green sleeved pvc sheathed cable) should be run from the earthing terminal of the

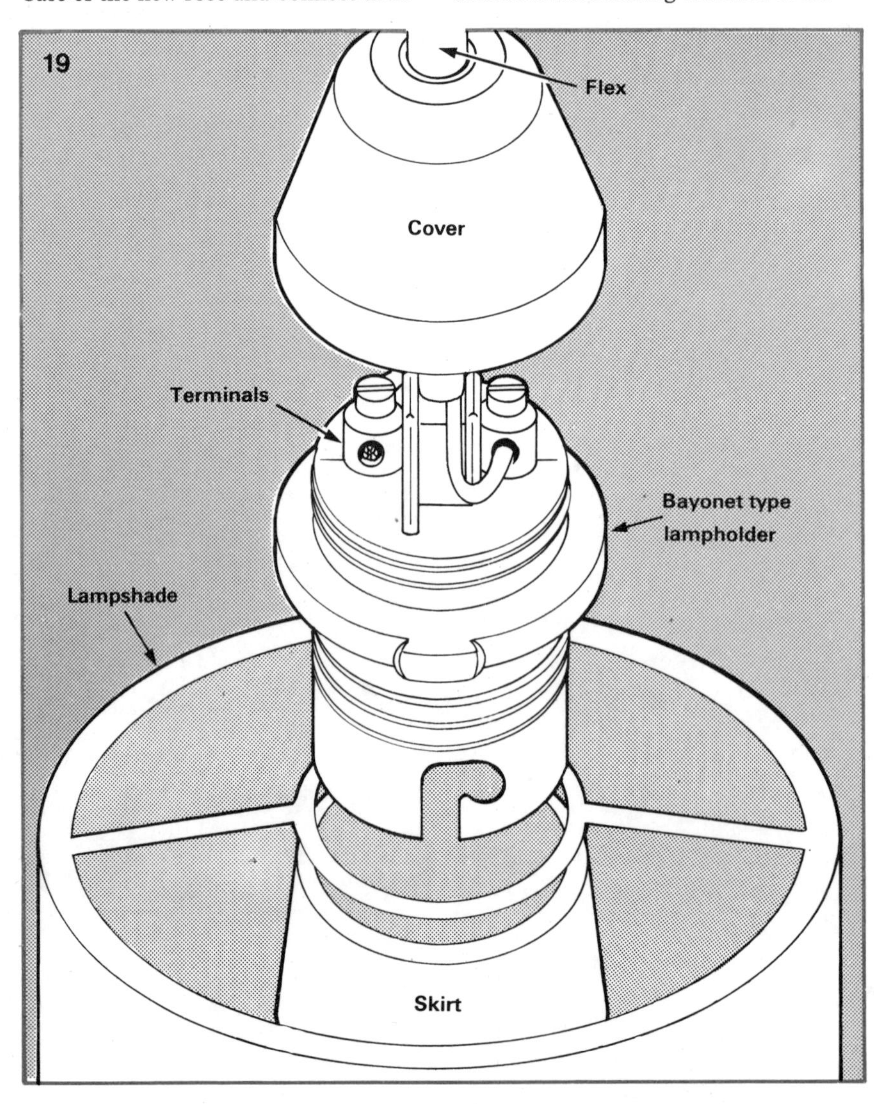

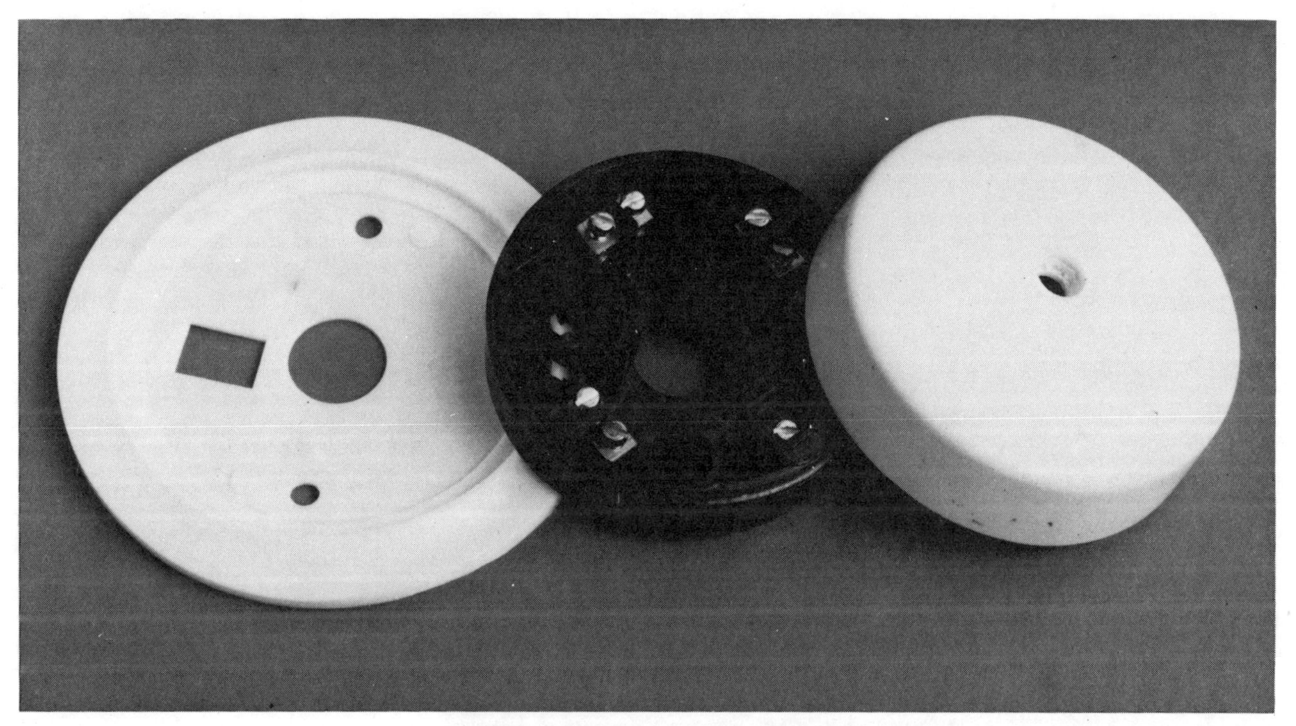

Fig. 20 The component parts of a ceiling rose

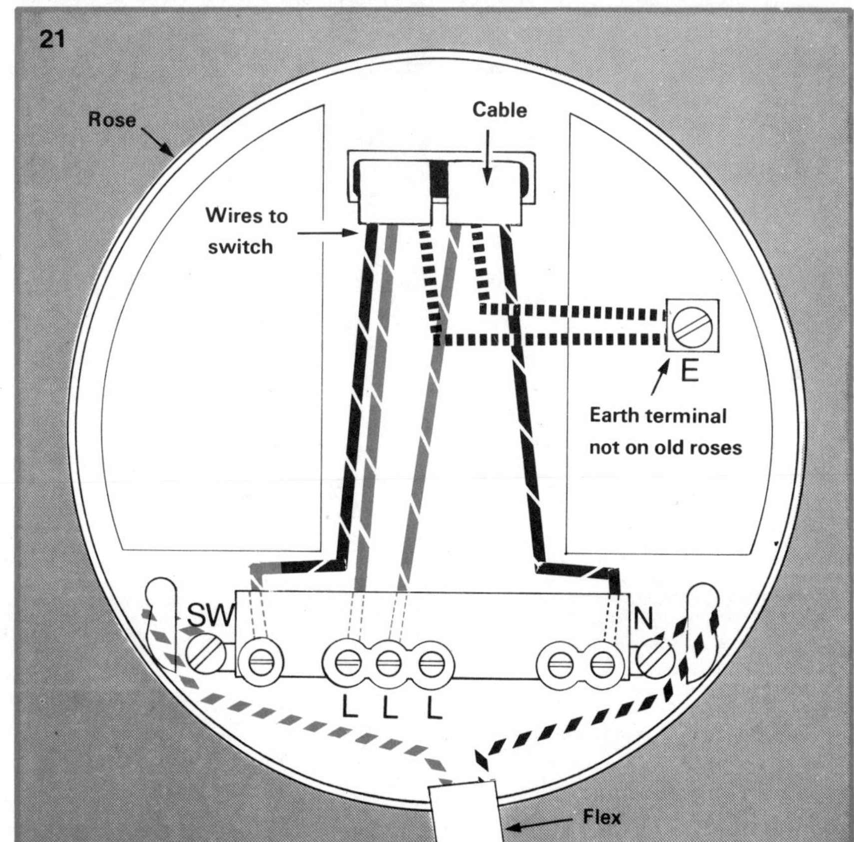

Fig. 19 Unscrew the cover to reveal a lampholder's flex terminals

Fig. 21 Electrical connections at one type of ceiling rose

new rose back to the earthing terminal strip on the consumer unit.

If, however, yours is an old installation which is likely to be completely rewired in the near future, this earth wire can be omitted until the work is done. But if the lighting fitting is a metal one, the earth wire must be fitted.

With all the cables fixed to the terminals securely, screw the base of the new rose to the ceiling. Fit the flex wires as described. Always choose a heat resistant flex if this has to be renewed.

TESTING APPLIANCES

Most of the electrical appliances sold today are reliably made and will give good service provided they are looked after properly. In most cases it is unwise to attempt what may turn out to be complicated repairs. This applies particularly to the larger appliances such as washing machines and refrigerators, but it is also unwise to meddle with smaller appliances unless you know exactly what it is you are doing.

It is true that many do-it-yourself enthusiasts do most, if not all their appliance repairs, but it is not a practice I would recommend for beginners.

When portable appliances stop operating for no apparent reason, however, there are some simple tests which can be made.

The first thing to do is to switch the appliance off at its own switch and at the switch at the socket outlet. If there is no switch at the outlet, pull out the plug. If the plug is a fused type, the fuse may have blown. This can be tested as shown in *fig. 11*.

If the fuse is in order, examine the flexible lead from the appliance. If this also is in order, try another appliance (which you know is in good order) in the socket outlet. If this works, there is something wrong with the original appliance and it should be sent for servicing.

LIFTING FLOORBOARDS

If you have to lift any floorboards to do a wiring job, you will need several tools: a hammer, a couple of electrician's bolster chisels, a cold chisel about 1 ft. long, a tenon saw, handsaw and a nail punch.

Look for a board which does not run the full length of the room, but butts up to another length of floorboard. Use the bolster chisels to prise the board up, *fig. 22*, and the batten chisel to keep the board in a raised position, *fig. 23*.

Move the bolsters along the board in order to ease it up. Get someone to raise the end of the board, then move the cold chisel along underneath it as far as possible.

Let go of the board and press its end downwards. The rest of the board should then come up, the cold chisel acting as a fulcrum.

Repeat the process until you can finally pull the floorboard from under the skirting board. (Skirtings are usually fixed after the floorboards have been laid.) Remove the nails from the floorboard and any still in the joists.

Fig. 22 Use a bolster chisel to prise up a floorboard

Fig. 23 Use a wooden batten to keep the free end of the board raised

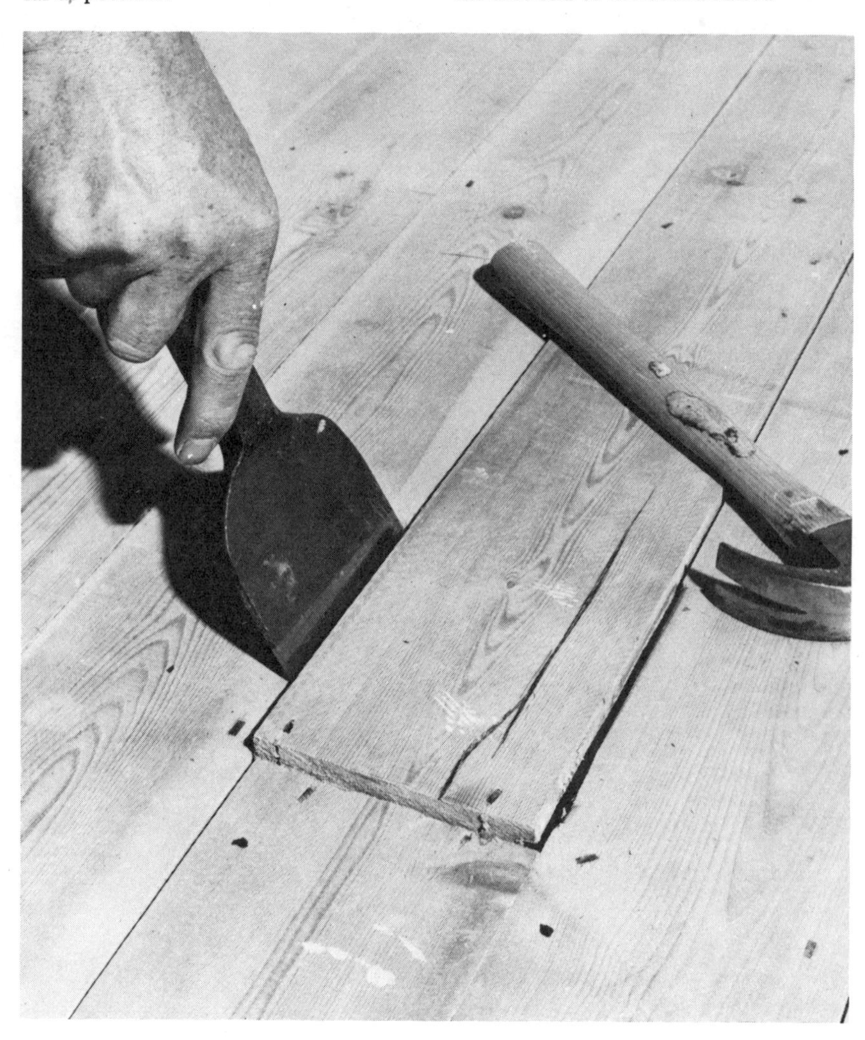

WATCH OUT FOR CABLES

Floorboards that run the entire length of the room will have to be cut. Great care must be taken when doing this so as not to cut through cables or gas and water pipes. Make sure, therefore, that the current is turned off at the main switch before you start.

Begin by drawing a line across the board in front of the nail holes which secure it to the joist at one end. You will now need a pad saw. Make a starter hole for the saw by drilling holes closely together to form a sufficiently wide gap into which the blade of the saw can be slipped. The thin blade is ideal for cutting across a floorboard since it is designed to work in confined areas. Insert a batten under the cut end of the board to keep it raised. Saw through the other end of the board on top of the joist. Then lift it clear. Support it, when refitted, by a batten, screwed firmly, to the joist.

T & G BOARDS

If the boards are tongued and grooved, the tongue on the board to be lifted, and that on the one next to it, will first have to be cut off with a handsaw. Here, too, special care should be taken

Fig. 24 Cut through the tongue of a tongued and grooved board with a pad or hand saw

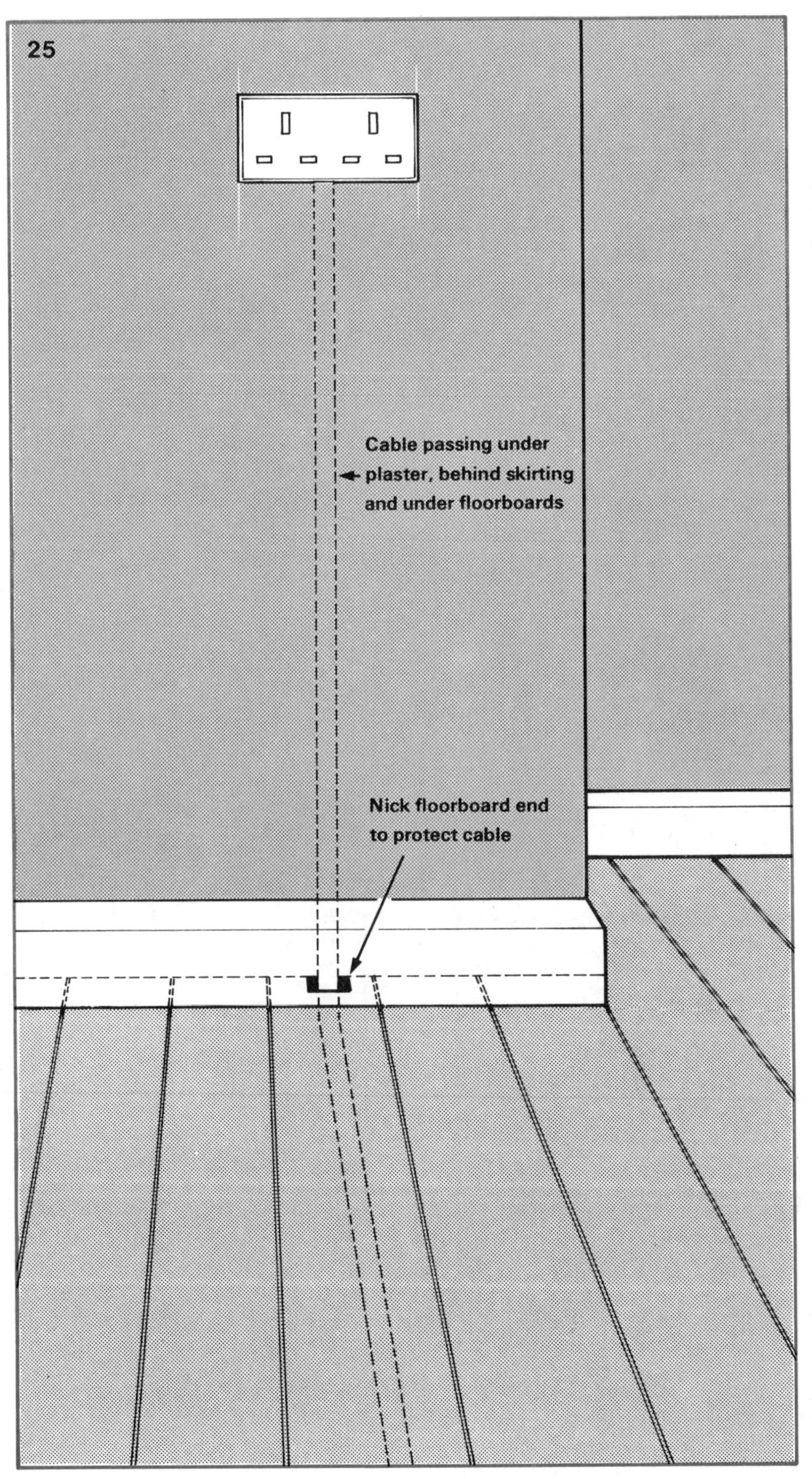

to avoid cutting pipes or cables.

It is a good idea to replace floor-boards with screws rather than nails so that the cables will be accessible.

One more tip. As floorboards usually extend under skirtings, any cable rising up the wall from the area under the boards could be damaged by the end of the board. This can be avoided by nicking the end of the board to form a small channel for the cable to pass through, *fig. 25.*

Fig. 25 Nick the ends of floorboards to protect cables

Fixing Electric Bells

Any handyman can tackle the job of fixing an electric bell. Nowadays you can buy all the necessary parts in a complete kit except, of course, the battery.

The simplest circuit, *fig. 26*, consists of a bell or buzzer which works when a bell push fixed at the front door is pushed; a suitable length of two-core insulated bell wire; a $4\frac{1}{2}$ volt battery; insulated staples to fix the wire; and fixing screws to fit the push and bell.

To install such a circuit, fix the bell in the selected spot. Drill a hole in the door frame for the wires which go to the back of the bell push. Before fixing the bell push to the frame of the door, fix the two ends of the bell wire to its terminals. One of these wires will need to be cut and its other end fixed to one terminal of the bell. The rest of that length of wire runs from the other bell terminal to a terminal on the battery (it does not matter which).

The other (uncut) length of wire from the bell push runs to the other terminal on the battery.

When the push is operated, the circuit will be complete. Current will flow through it and the bell will ring.

The battery can be housed in a home-made wooden box positioned near the bell or stood on a convenient shelf – perhaps in a cupboard. You can, of course, buy bells which themselves house the battery, thus reducing the amount of wiring which will be visible.

The bell wire can be anchored to skirtings and door frames using insulated staples.

EXTENSION BELL

If a bell is clearly audible in one part of the house but not in another, an extension bell can be fitted to overcome this problem. *Figures 27a* and *27b* show how two bells can be wired up.

For the best results, the bells should be similar types (i.e., their characteristics should be the same). Too long a wire to the extension bell should be avoided as this will mean

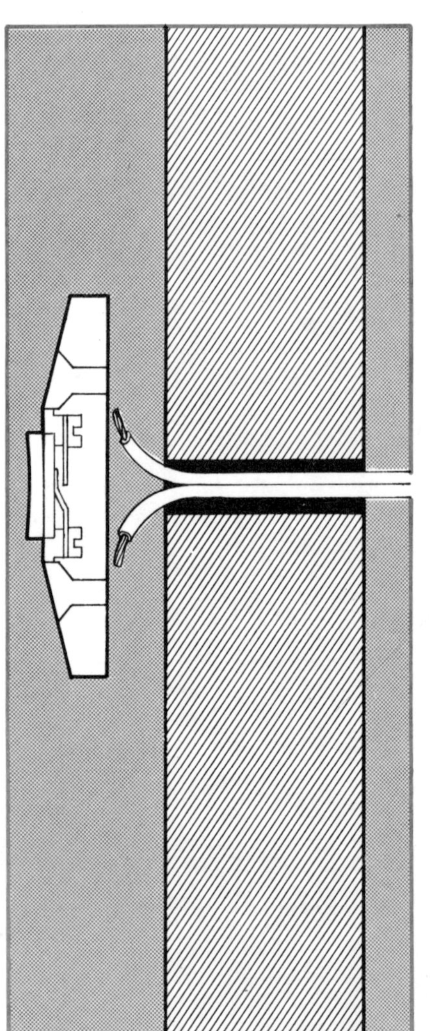

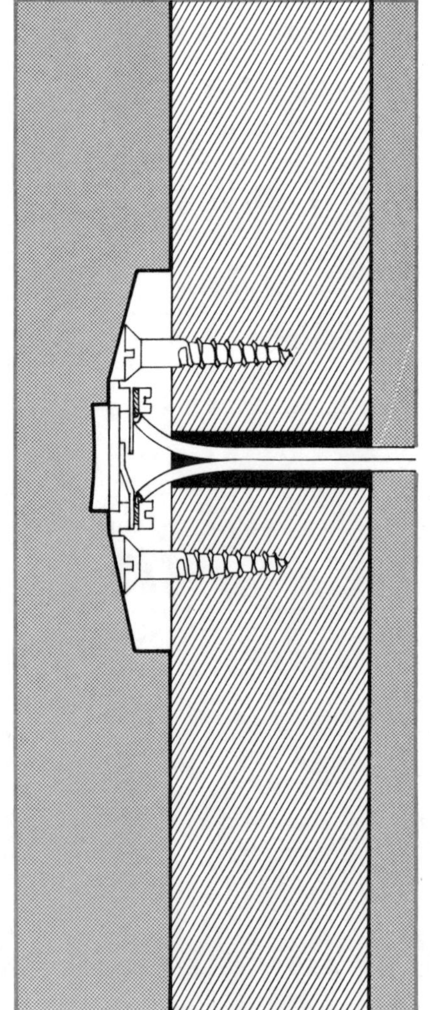

Fig. 26 Simple battery bell circuit

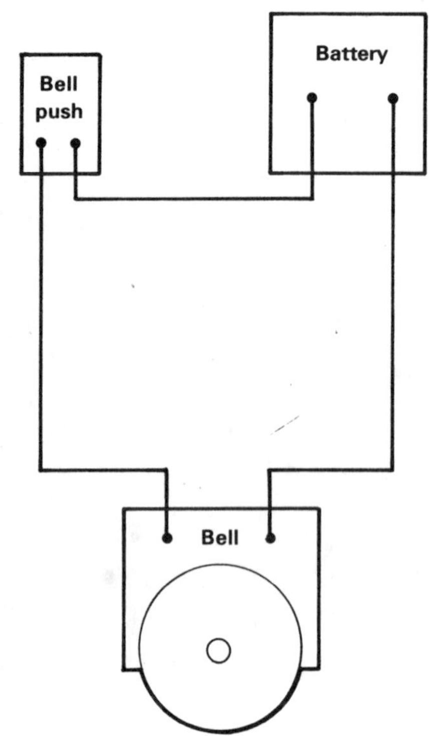

its resistance to the current will be high.

In *fig. 27a*, the bells are wired in series. In *fig. 27b* they are wired in parallel.

No detailed explanation of the difference between parallel and series wiring is necessary. All you need to do is to follow the diagrams.

Some experimenting, however, may be necessary to get both the bells to work satisfactorily, so if one method of wiring will not work, try the other.

Problems sometimes arise with the parallel method of connection. One is that one of the bells may cause the amount of current supplied to the other to be reduced. The answer to this one is to increase the voltage slightly with a stronger battery or to adjust the contact setting screw of the bell. The position of this screw may vary according to the make of bell, but *fig. 28* showing the contact screw of an ordinary trembler type of bell will give you a clue.

If an extra bell is wired in series you may again require a stronger battery to drive the extra current needed around the circuit. A suitable type of battery for this purpose is a six volt lantern battery.

I suggest that you experiment first with both methods of connecting before you finally fix the bell and the wire in position.

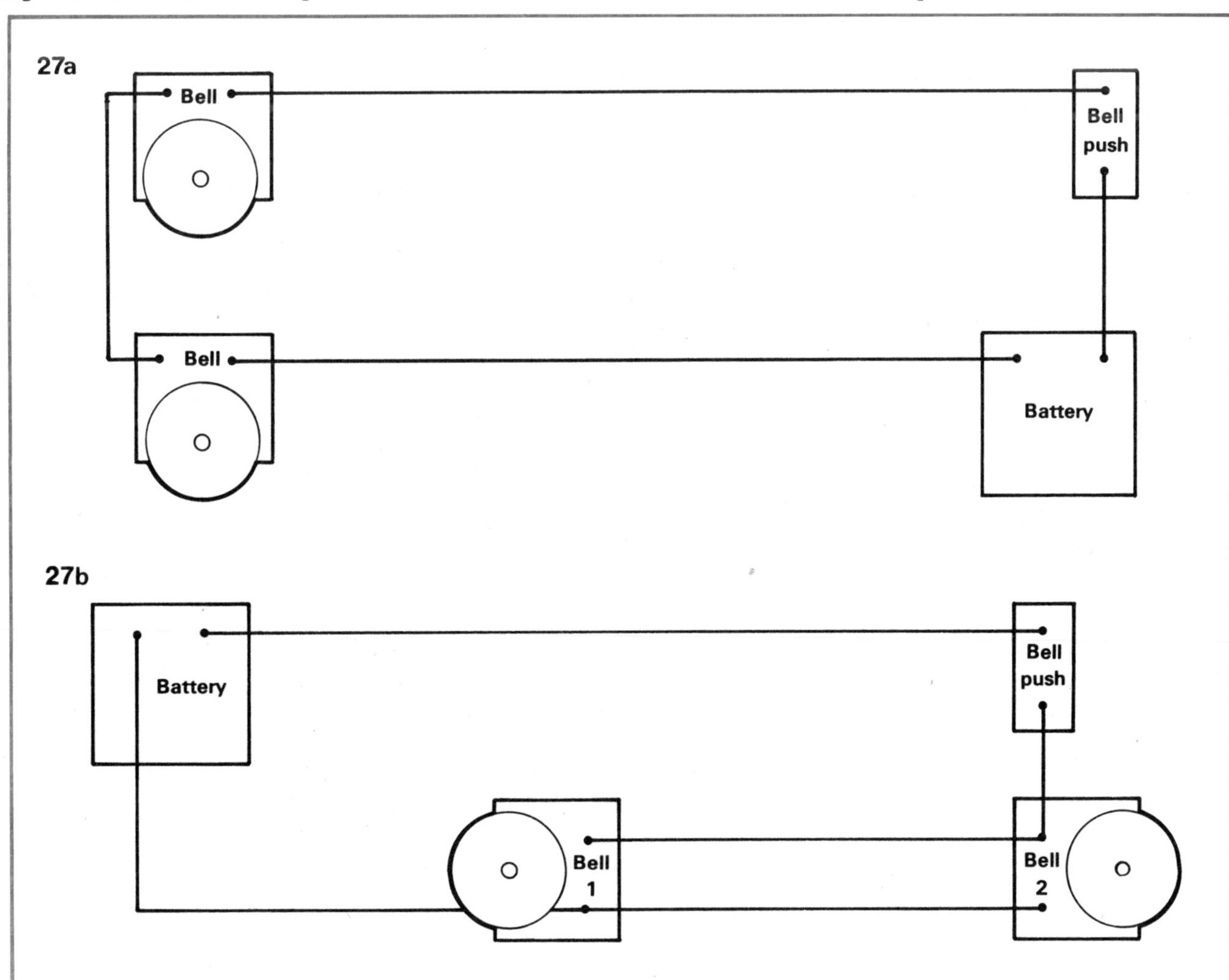

Fig. 27a Bell circuit wired in series

Fig. 27b Bell circuit wired in parallel

OFF THE MAINS

For operating door bells or chimes from the mains a double-wound transformer with its secondary winding earthed is essential. Transformers sold especially for bell circuits incorporate this safety device as a built-in feature. Make sure this is the type you buy.

There are two windings to a transformer – the primary and the secondary. The primary (input) is connected to the mains, and the secondary (output) to low voltage appliances such as bells, *fig. 29*. Transformers sold for operating bells or chimes have various tappings on the output side. These enable you to choose the voltage you require.

The tappings may be four, eight and 12 volts *or* three, five and eight volts, according to type. The most suitable should be chosen after experimenting.

WIRING UP

To wire up a bell or chimes from the mains you need two-core and earth pvc-sheathed house wiring cable, size 1·0 mm.² or 1·5 mm.². Position the transformer near the consumer unit. Run a length of the cable from a spare 5 amp fuseway in the consumer unit to the primary terminals of the transformer.

Then run a length of bell wire from the secondary terminals of the transformer to the bell push and to the bell or chimes (as for the battery operated circuit already described.)

If there is no fuseway to spare in the consumer unit, or this method of wiring is not convenient, there are other ways of wiring bells from the mains. These are described on page 71.

Fig. 28 A contact screw of a bell is adjustable

Fig. 29 Transformer connections for bell circuit

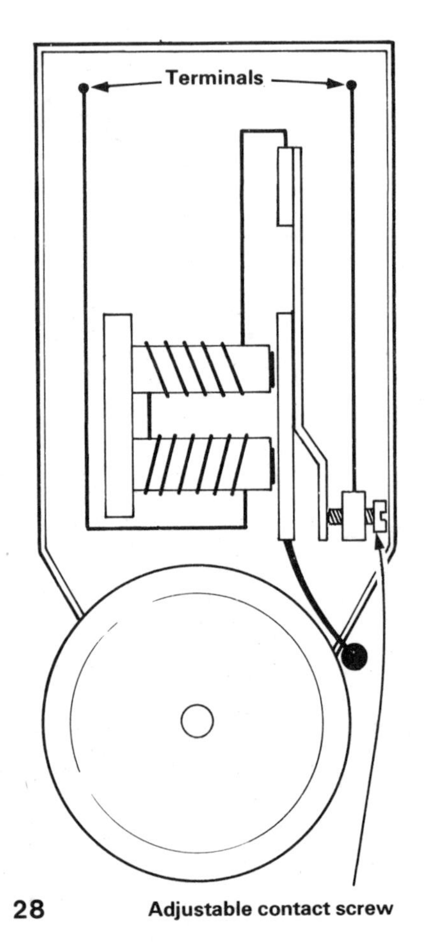

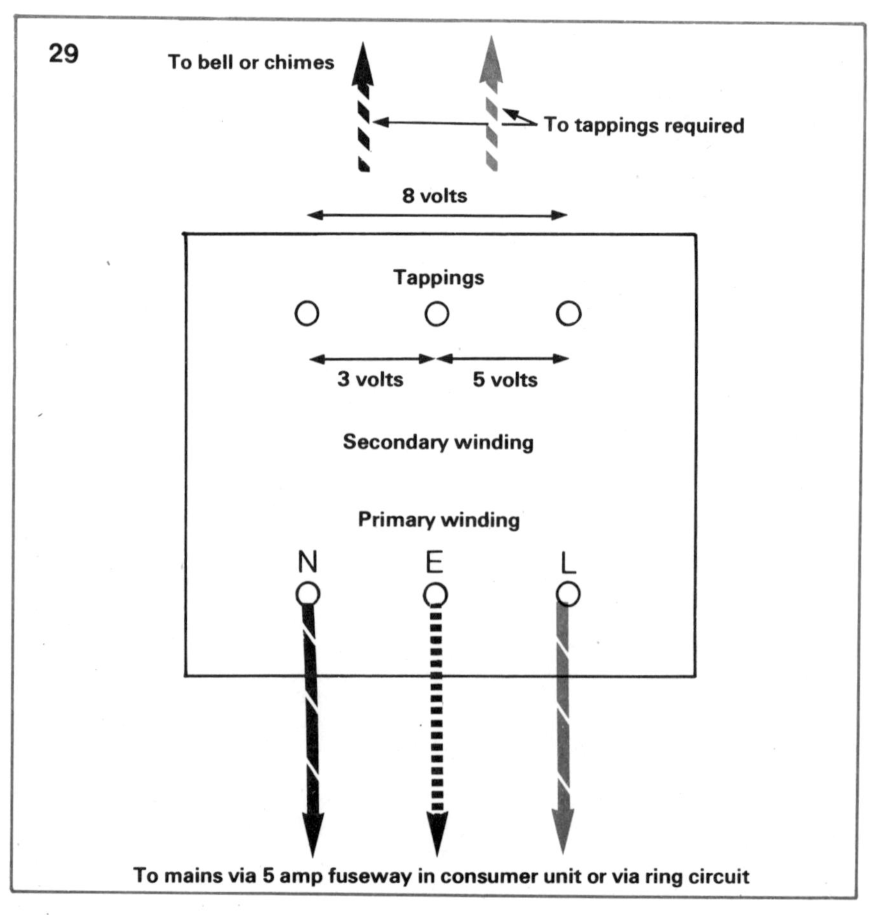

Watts, Volts and Amps

There are three terms in particular used in the electrical trade with which householders should become familiar.

Watts is the term heard most often, perhaps. This is the measurement of the amount of electricity used at a given moment by an appliance.

Volts, the second term, represent the measurement of pressure which forces current to flow along a conductor.

The amount of flow of electricity along a conductor is measured in *amperes*, usually abbreviated to amps.

Earlier on, I said it was important to know the wattage of an appliance. This is usually indicated on a metal plate or label on the appliance – it can often be found on the back. This may appear, for example, as 2,000 watts (2 kilowatts).

On the same label, the voltage of the appliance is sometimes marked also. This will almost certainly be 240 volts, as this is now the standard voltage in most parts of Britain.

SIMPLE DIVISION

It is sometimes helpful to know how many amps a plug, cable or socket outlet can cope with, and you can find this out easily if you know the wattage of the appliance and the voltage of the mains supply.

All you have to do is to divide the wattage by the voltage. For example, to find out the amperage of a plug required for a 2,400 watt appliance, divide 2,400 by 240. The result is ten amps.

In this case a 13 amp rectangular-pin plug is suitable – the type used in socket outlets on a ring circuit. The plug should, in this case, be fitted with a 13 amp fuse.

If the wattage of an appliance is, say, only 480 watts, only a 3 amp fuse needs to be fitted in the plug: 480 divided by 240 volts equals two amps.

It will be seen that the higher the wattage, the greater will be the power expended by the appliance.

RESISTANCE

There is one other term widely used by electricians – resistance. All appliances and circuits offer resistance to the flow of electricity. The resistance of an appliance or cable is measured in ohms.

The three units – volts, ohms and amps – are co-related. If, therefore, any two of them are known, the third can easily be calculated by the use of three simple formulae:

Volts = amps × ohms.

Ohms = volts divided by amps.

Amps = volts divided by ohms.

Fig. 30 A typical voltage plate for an electrical appliance

Plugs and Connections

Most of the plugs used in domestic wiring systems nowadays have three pins. Two-pin plugs are still used but only for appliances which are double insulated and for some lighting fittings. The two-pin type of plug, *fig. 31*, of course, has no earth terminal.

All appliances which are sold with a three-core flexible lead should be connected to a three-pin plug.

Three-pin plugs are made in two types. First, the modern plug which is fitted with a fuse and has rectangular (flat) pins, *fig. 32*. These are all made in a 13 amp size, but 3 amp as well as 13 amp fuses can be fitted into them.

Round pin plugs, *fig. 31*, are produced in three sizes – 15, 5 and 2 amps. These are *not* fitted with fuses. In installations where these plugs are used, the circuit fuse in the consumer unit is the only fuse for the circuit to which the plug is connected.

EXTENSIONS

There are various connectors which can be used to fit an extension lead to an appliance. If you want to use an

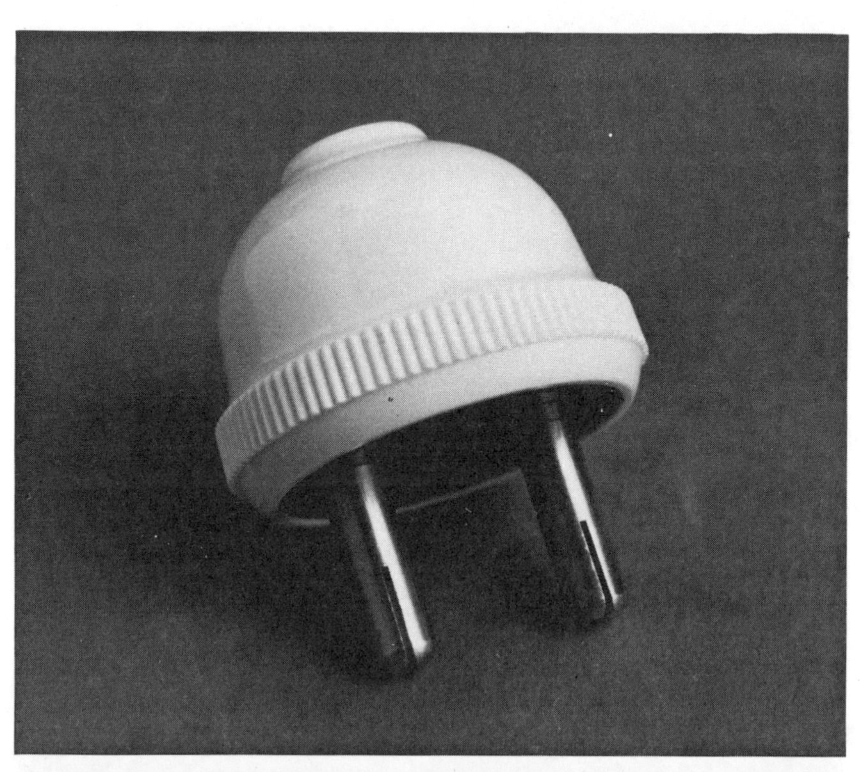

Fig. 31 Two-pin non-fused plug

Fig. 32 Modern 13 amp fused plug

extension flex with an appliance fitted with a three-pin plug, the flex has to be connected to a socket at one end and to a plug at the other. Sockets are available in 15, 13 and 5 amp sizes.

One way to increase the length of a flexible lead is to use a flex connector (not to be confused with a block connector). This has a shrouded socket and a non-reversible three-pin plug, *fig. 33*. There are also connectors with only two pins, *fig. 34*. These *are* reversible, which means that the wires can be connected to either of its terminals in safety.

CORRECT CONNECTION

The socket section of a flex connector should always be fitted to the lead from the socket outlet or bulbholder. The plug section of the connector should be connected to the lead from the appliance.

This is most important because if it is not done, the plug pins will be live if the connector is accidentally pulled apart while the plug is still switched on and therefore dangerous.

It is important, too, that all connections in these accessories are firmly made. Should the connector or

Fig. 33 Flex connector with non-reversible plug

Fig. 34 Other types of flex connector

the plug become warm, make certain that there are no loose connections at the terminals. If, after checking, you find that the plug still gets hot, it could be overloaded, so check that the plug is the right type for the purpose.

Never use a connector or a plug which is chipped or cracked. Replace it as soon as possible.

EARTHING APPLIANCES

Every electrical appliance must be adequately earthed unless it is double insulated. Appliances which are double insulated normally carry a label in the box to this effect.

If you buy a recognised make of appliance, there is little danger that it will not be thoroughly insulated if it claims to be double insulated, but always make sure that it does.

There is an easy way to check the safety of this type of appliance if it has a metal casing. All you need are a torch battery and bulb, and a couple of crocodile clips. Wire these up as shown in *fig. 35*.

When you want to test the earthing efficiency of an appliance, disconnect it from the mains. Fit one clip to, or hold it to, the metal of the appliance. Fix the other clip to the earth pin of the plug. If the appliance is adequately earthed, the bulb will glow brightly.

If there is only a dim glow from the bulb, and the battery is known to be in order, the appliance is inadequately earthed. There is, however, always a chance that there may be a fracture in the flexible lead, so check this.

As a further check, touch the clip to each of the other two pins of the plug in turn. There should be no light from the bulb. If there is light, the appliance is dangerous and certainly needs expert attention.

You can, of course, dispense with the crocodile clips and simply hold one wire to the appliance and the other to the pin on the plug.

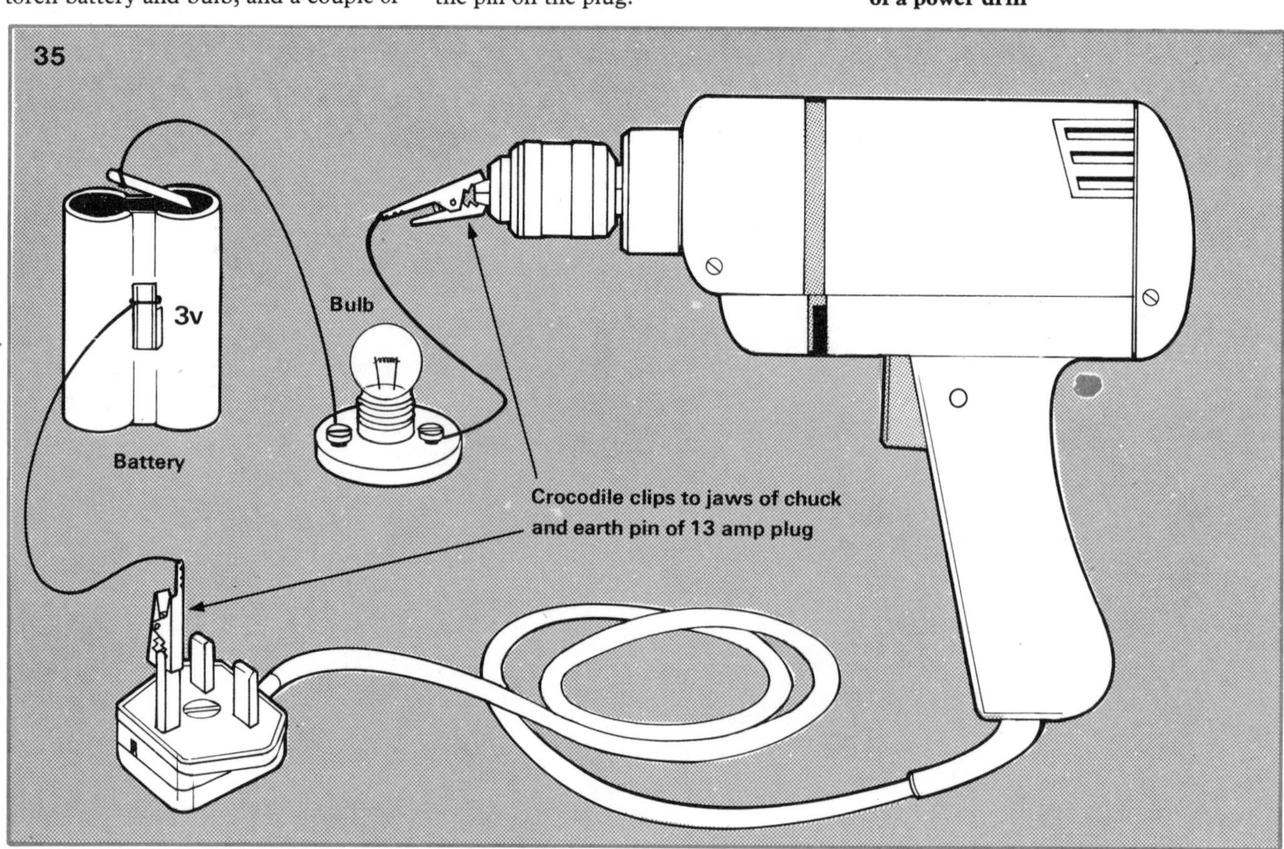

Fig. 35 Testing the earthing efficiency of a power drill

Fluorescent Lamps

Tubular fluorescent lamps have a longer life than ordinary lamps of the same wattage. Another of their advantages is that they cost less to run. They are available in various lengths and wattages.

Their length can range from as little as 6 in. to 8 ft. and their wattages from four to 125.

When choosing fluorescent lamps, allow ten watts per square metre, or about one watt to a square foot, for the floor area to be illuminated.

These lamps normally have a life of at least 5,000 hours. They should be replaced when the light becomes dim, when they are difficult to start or when their ends begin to become black, *fig. 36*.

Fluorescent lamps should not be allowed to flicker; this can damage the controls. Also, avoid switching them off unnecessarily. It is better to allow them to stay on for brief periods, such as when leaving the room for a few minutes, because each time they are switched on their electrodes shed some of their oxide coating.

As the majority of fluorescent lamps are fitted with metal casings, they should be earthed. Many houses, of course, will not have an earth wire in the cable running to the lighting point. In these cases, an earth wire should be run from the lighting point back to the earthing terminal strip of the consumer unit.

There are two types of fluorescent lamp. One works in conjunction with a starter switch and a choke; the other has no starter switch and is known as a quick start or instant start type.

The advantage of those with a switch start is that they last longer, but if you want immediate light, then the quick start type is the one to choose.

Another advantage of the quick start lamp is that it is noiseless. With the other type, there is a slight hum from the choke. One possible cure for this is to have the choke fitted away from the tube – perhaps on a joist and enclosed in a box.

FAULTS

The odd thing about fluorescent lamps is that they rarely fail suddenly or entirely. And provided the fault is in the tube itself and not in a component, some sort of life is apparent in the lamp.

There are, however, a number of things which can go wrong with this type of lamp.

For example, if a lamp seems to be dead, the circuit fuse or the fuse in the fitting may have blown; there could be a break in the circuit wiring somewhere; or, in the case of a switch-start type, the lampholder may be at fault (i.e., faulty contact).

If the lamp refuses to start when switched on, you may notice that the electrodes are glowing at each end. The colour of the glow can give you a clue to the fault.

If the electrodes glow white on a switch start tube, the starter is probably faulty. If they glow red, then the tube has reached the end of its useful life and should be replaced.

This sort of fault on a quick start type of tube may be due to an inadequate earth connection, or the wrong type of tube may have been fitted.

Sometimes a tube will make a number of efforts to start but fail to do so. On a single tube this could be due to a fault in the starter. If the model is a twin tube type, the lampholder connections may be crossed.

Sometimes, when a new tube is fitted, it will light but with a shimmering effect. This is usually nothing to worry about; the tube will soon settle down and work normally.

There are a number of possible faults which can cause a lamp to flicker on and off. It could be due to low mains voltage, a faulty starter switch, the lamp itself may be faulty, or it may have reached the end of its useful life.

If the flicker is slow and very noticeable, then the lamp needs renewing.

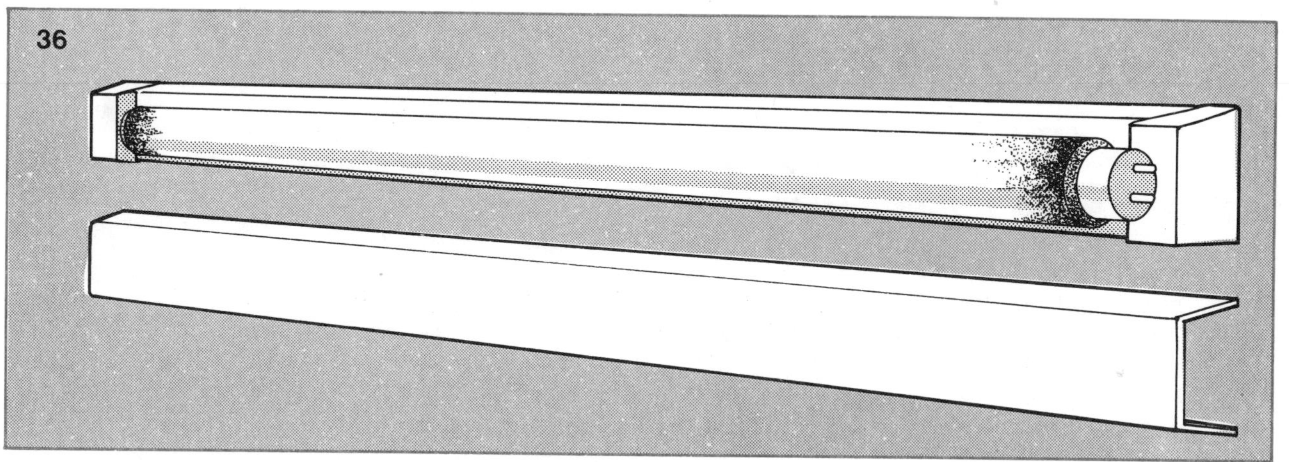

Fig. 36 Fluorescent lamps should be replaced when the ends turn black

Caring for Heaters

To get the best results from an electric heater, the appliance needs regular maintenance. Basically, there are two ways in which the heating element of a radiant type heater is mounted.

Reflector types, *fig. 37*, have a thin element wound around a rod or tube. This element is fitted in front of a metal reflector which reflects the heat back into the room.

In a fire bar type of heater, the element wire is in the form of a coil set in the grooves of a slab of fire clay, *fig. 38*.

All fires of this type should be fitted with a wire guard in front of the element.

An electric heater should never be connected to a lighting point or to a plug which has only two pins.

Take care always that the flexible lead does not cross in front of the heater.

Reflectors should be cleaned regularly with metal polish, not with the scouring type of polish or powder. The heater should be unplugged from its socket outlet before it is cleaned and must be completely dry before being used again.

Convector-type heaters are prone to gather dust around their elements. Some types can be cleaned by turning them upside down and using a vacuum

Fig. 37 Radiant heater, reflector-type

Fig. 38 Fire bar radiant heater

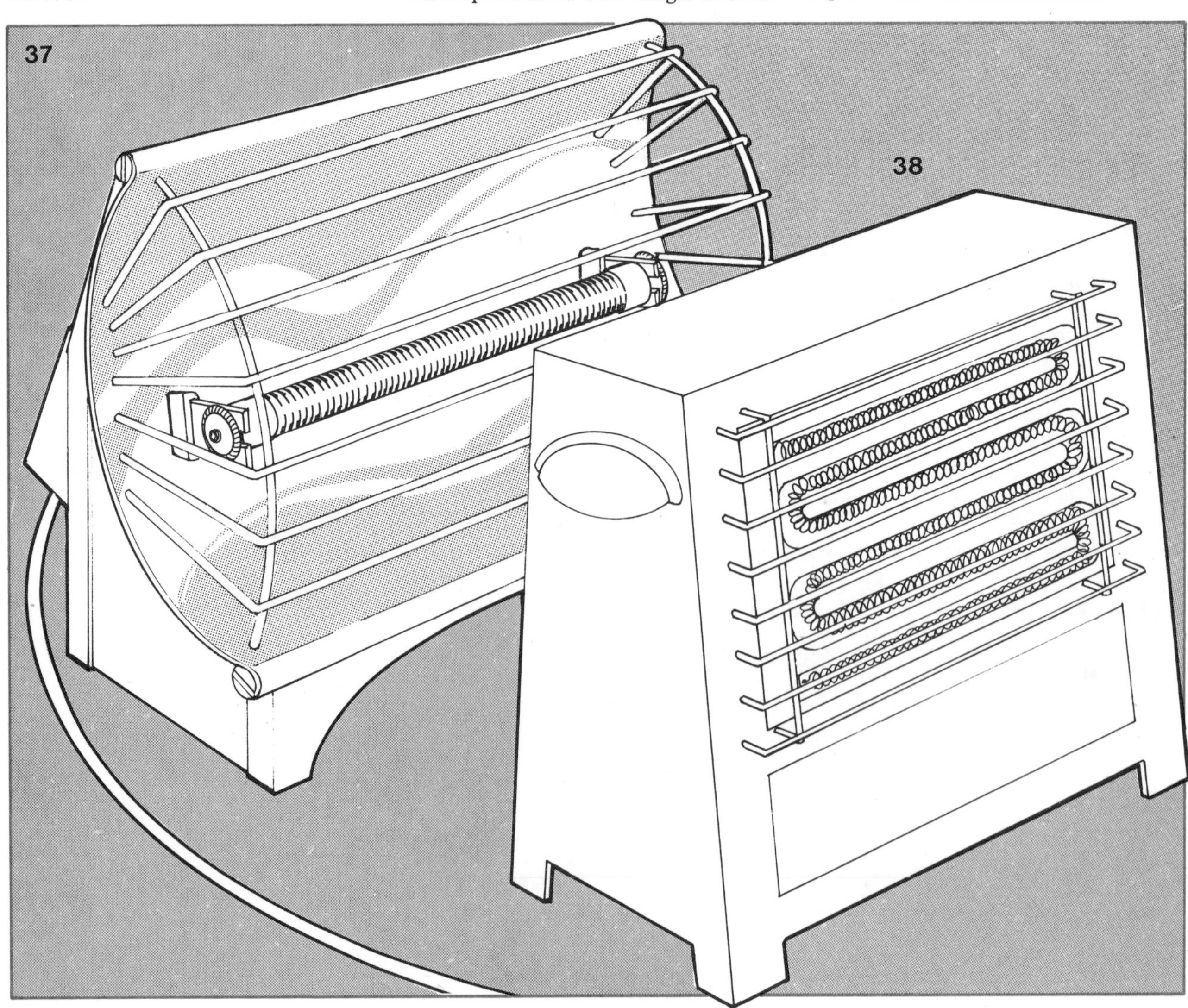

cleaner. But be very careful, if you do this, not to damage the heating element.

SIMPLE REPAIRS

With an ordinary reflector type of heater there is not a great deal which can go wrong. A damaged or broken flex lead is one common fault and that is easily rectified.

A more common fault, perhaps, is the failure of the element. These are usually the pencil rod types, *fig. 39*, and may be fitted to the fire in three ways: they may simply fit into spring clips; they may be held by nuts; or they may have dagger-type end caps which are fitted into spring-loaded contacts.

To replace a pencil rod element, first remove the guard. This can be done easily as a rule by pressure from the fingers. Then undo the screws which hold the shields covering the ends of the element. Release the element from its fixture. The way this is done will depend upon the make of heater.

Put your new element in (and do make sure you have the correct type), screw back the end shields and replace the guard.

Most other types of electric heater have their working parts enclosed and are best left to the expert for repairs.

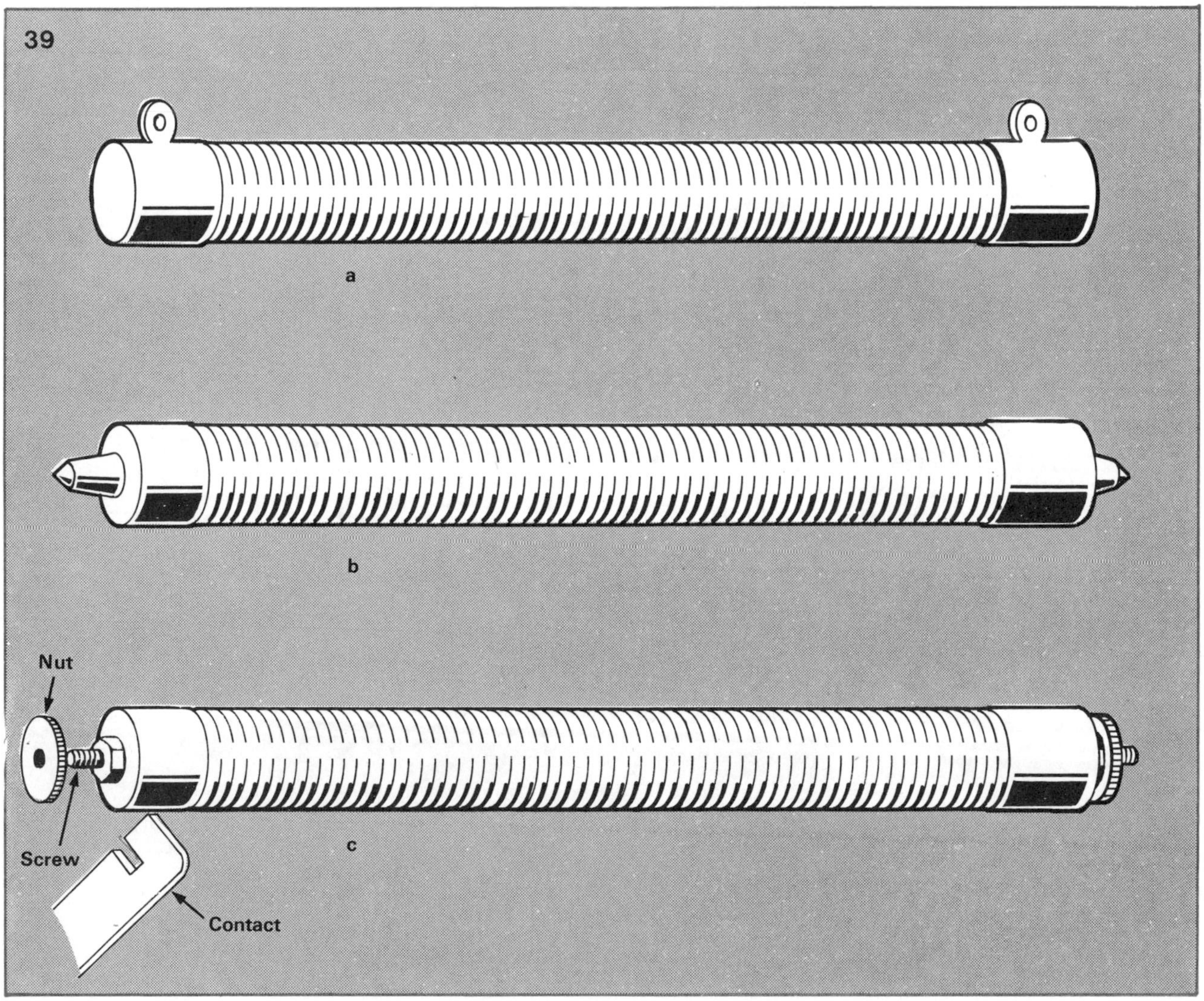

Fig. 39 There are various types of electric fire elements. Make sure you buy a replacement similar to the one you are discarding

39

a

b

Nut

Screw

c

Contact

Ring Circuits

Before ring circuits came into being, every power point, as it used to be called, in a domestic wiring system had its own circuit.

In an old installation, therefore, this meant that a house with, say, six power points had six power (heating) circuits originating at the fuseboard!

All these individual circuits were in addition to separate circuits for lighting, cookers and immersion and other types of water heaters. The result – confusion at the fuseboard!

This arrangement also meant that if the occupier of the house decided that extra power points were necessary, separate circuits had to be installed for each.

Some houses are still wired in this way, using round pin socket outlets, and having an old-fashioned fuseboard or a multiplicity of fuse units instead of a modern consumer unit.

My advice to anyone who has a confusion of wires and fuse units such as this is to get the house rewired with modern equipment so that all the circuits originate in one neat consumer unit.

Many householders have installed their own ring circuits; others have rewired their homes completely. But to do either of these things, you need to be familiar with all the necessary cables, accessories and circuits before you begin.

Fig. 40 Confusion at the fuseboard with an old type installation!

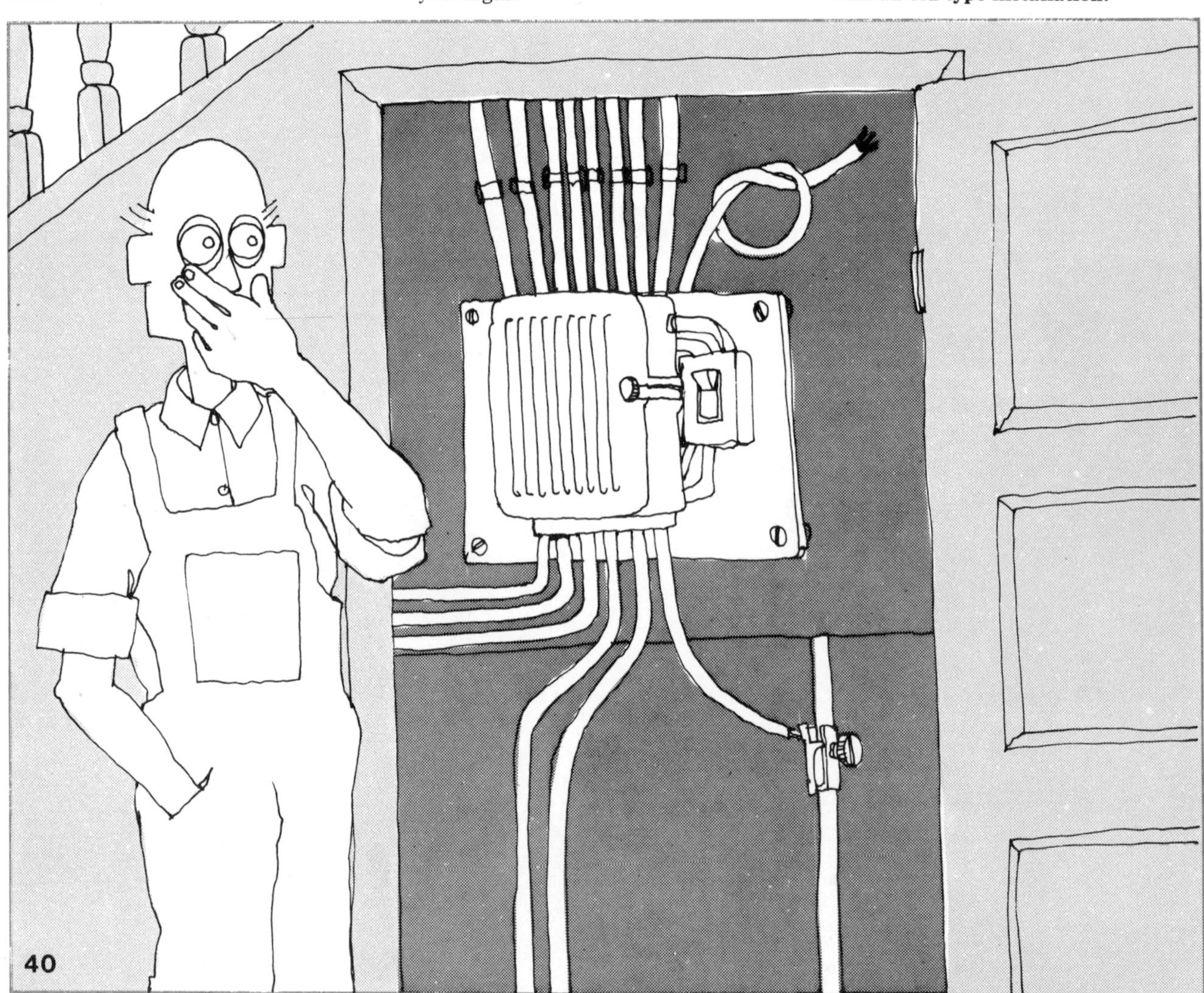

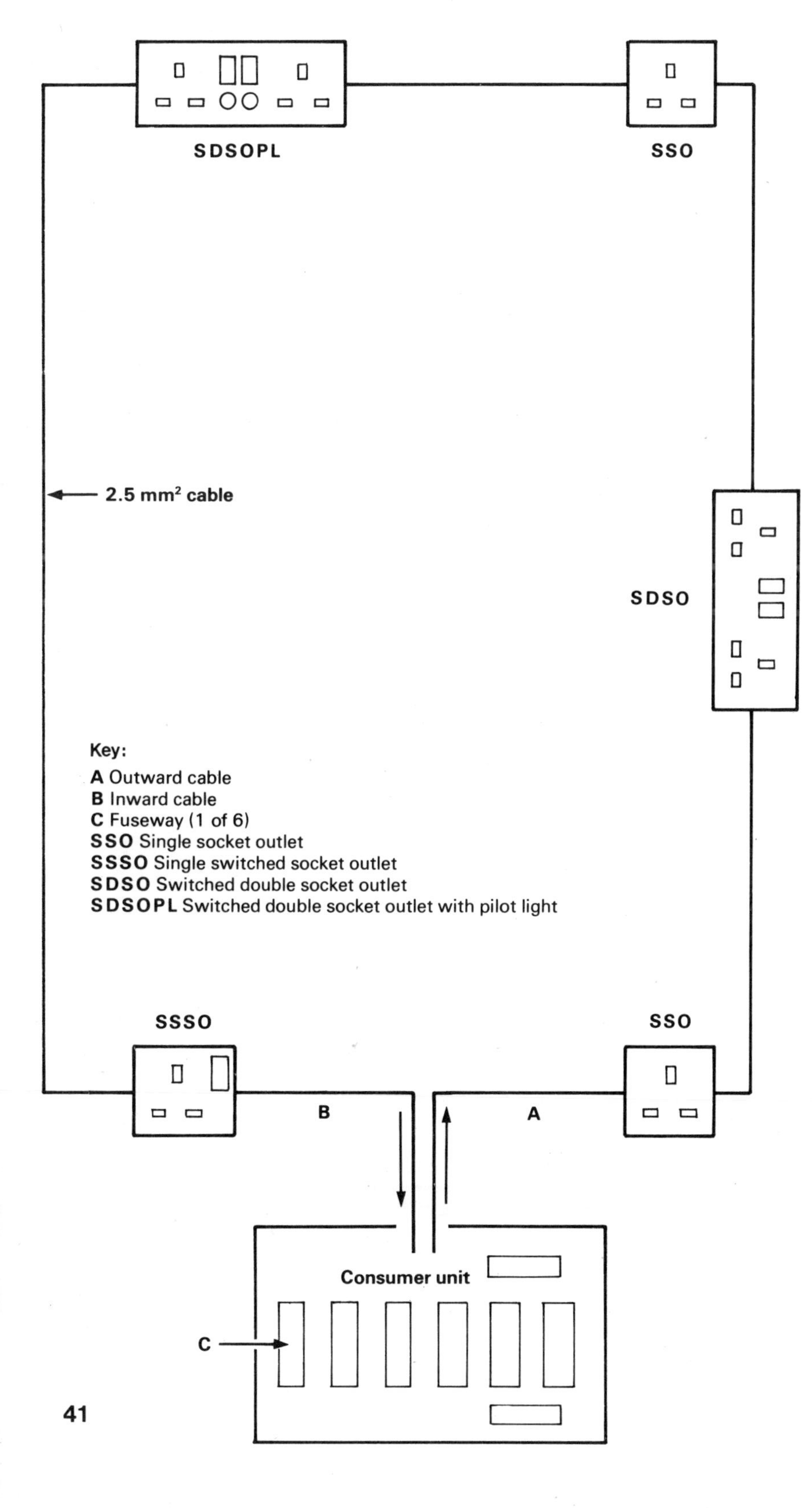

Describing such operations in complete detail is beyond the scope of this book; but I hope that the explanation of circuits given here, plus hints on various wiring jobs, will help readers to understand how a ring circuit installation works. Once this understanding is complete, carrying out the installation job itself becomes a practical possibility.

BASIC CIRCUIT

First, let me describe a basic ring circuit, *fig. 41*. A ring circuit is so-called because its cables, in effect if not in design, follow a circular route.

The cable starts its journey from the consumer unit and eventually it returns to it. On its journey throughout the house it is looped into and out of all the 13 amp socket outlets connected to the circuit.

At the consumer unit, the live (red) conductor of the cable is fitted to a fuseway terminal; the neutral (black) conductor to the neutral terminal strip; and the earth conductor (bare wire fitted with a green sleeve) to the earthing terminal strip.

Fig. 41 Basic ring circuit

Figure 42 illustrates how our basic ring circuit shown in *fig. 41* is wired up.

After the cable has completed its journey to all the socket outlets en route, its three conductors return to the consumer unit. There the live conductor is fitted to the same fuseway from which it started its journey. The neutral conductor returns to the neutral terminal strip and the earth conductor to the earthing terminal strip.

If you refer back to *fig. 4* page 7 and what I said earlier, you will now see why more than one wire in the consumer unit is connected to a terminal.

Virtually an unlimited number of socket outlets can be connected to a ring circuit though the number which can be used at the same time will be limited. The number shown in *fig. 41* has been limited for the sake of simplicity.

Provided the area of the floor (ground or first) where the ring circuit is to be installed is not greater than 1,080 square feet, only one ring circuit will be needed. If, however, the area is greater than that, an extra ring circuit will be necessary for each additional 1,080 square feet or part of that figure.

CAPACITY

One ring circuit is usually enough to suit requirements for small houses and flats. However, it is better to install two circuits in houses, or large bungalows, because this will double the capacity of the installation.

For example, a single ring circuit has a capacity of about seven kilowatts. This means that no more than three two kilowatt (2,000 watts) electric heaters and one kilowatt (1,000 watts) heater – or their equivalent – can be in use at the same time. If other appliances are switched on while these are in use,

Fig. 42 Connections for basic ring circuit in fig. 41

Fig. 43 Basic ring circuit with two spurs

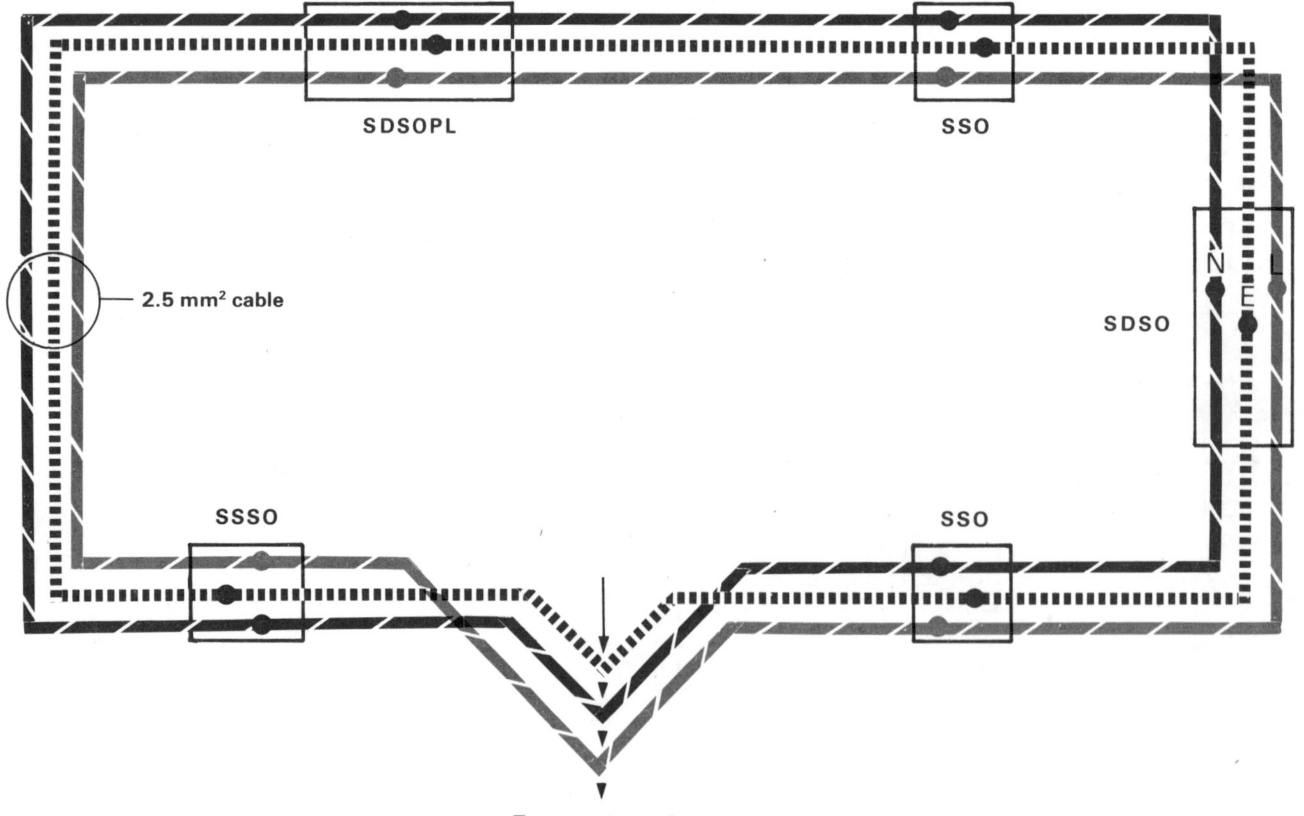

42

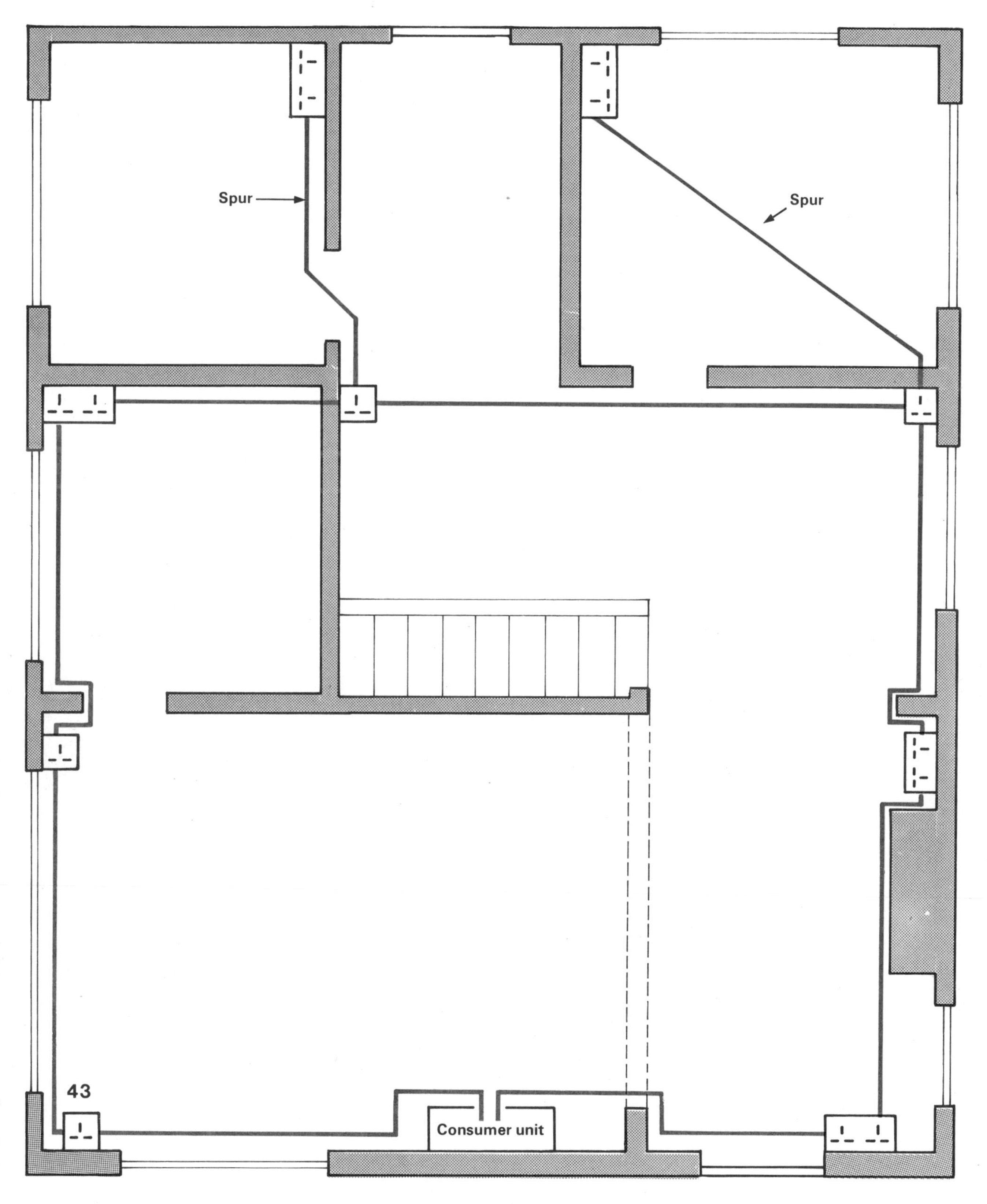
Spur
Spur
43
Consumer unit

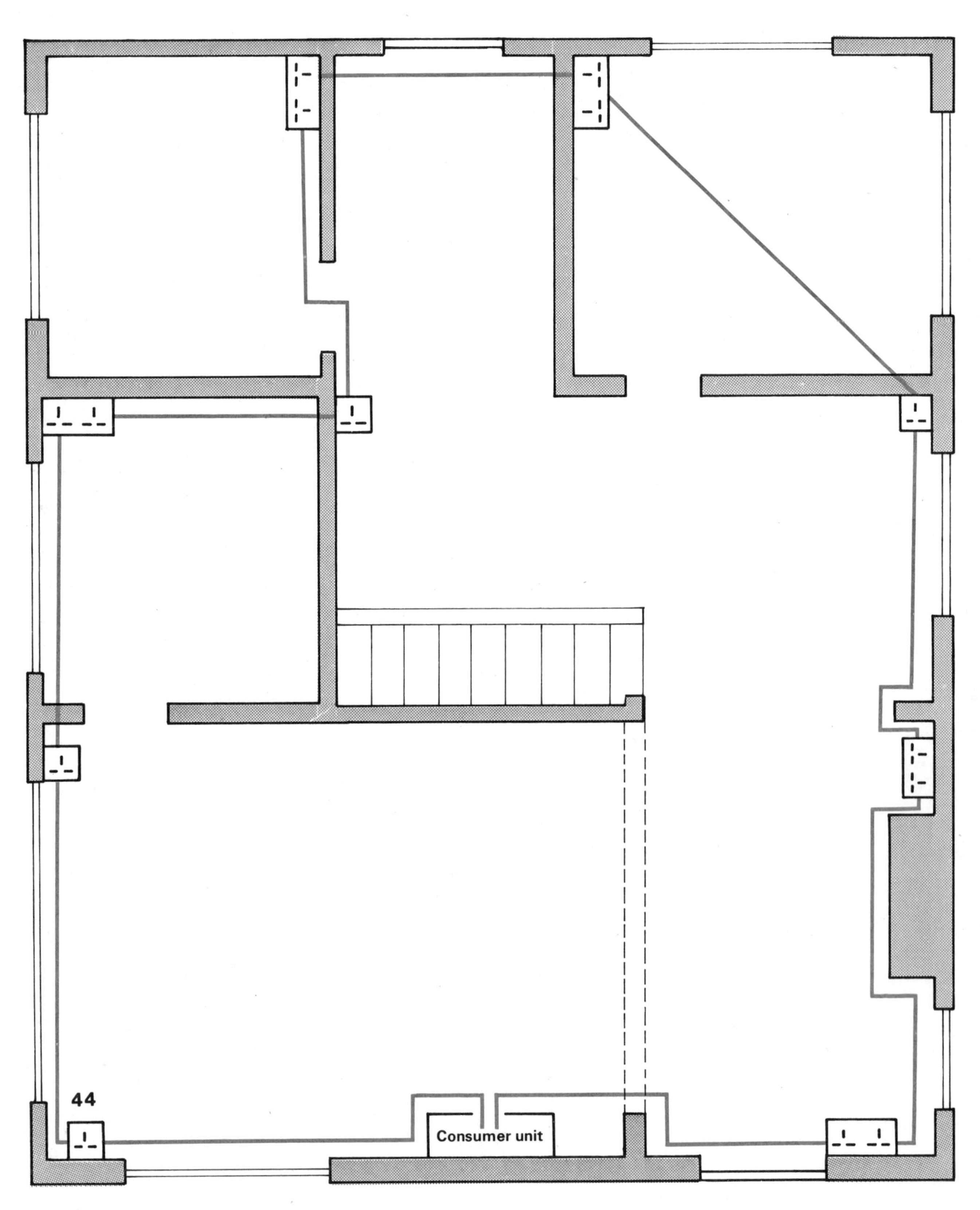
44
Consumer unit

the circuit may become overloaded and the circuit fuse in the consumer unit will blow.

The total capacity of two separate ring circuits, therefore, is roughly 14 kilowatts. This means, of course, that seven two kilowatt heaters (or their equivalent) can be in use at the same time, *provided that this load is evenly distributed between the two circuits*. Remember they are separate circuits, each fitted with its own individual circuit fuse.

Many houses are wired for two ring circuits – normally one for the ground floor and another for the first floor. This method of allocating one circuit to each floor makes it easy to remember which socket outlets are on each circuit, thus reducing the possibility that trouble may be caused by overloading either of them.

SPURS

Most ring circuits have a number of spurs connected to them. A spur is a single cable (a branch, if you like) which feeds socket outlets situated in remote parts of the house; for example, in the hall, in the far corner of a room – or in a garage attached to the house.

The use of spurs can save cable. In *fig. 43* a circuit is illustrated using two spurs. *Figure 44* shows a similar circuit wired without spurs.

Each spur can feed only two single socket outlets *or* one double (twin) socket outlet (*figs 45* and *46*). Not more than half the total number of socket outlets on the ring circuit *and* on the spurs can be fed from spurs. If the total number of socket outlets (including those on spurs) is 28, for example, no more than 14 of these should be fed from spurs.

Fig. 44 Basic ring circuit; no spurs

Fig. 45 Single wall-mounted unswitched socket outlet

Fig. 46 Twin wall mounted switched socket outlet

CONNECTING SPURS

There are various ways in which spurs can be connected to a ring circuit. They can be taken direct from the terminals of a socket outlet; or a joint box can be inserted in the ring cable, *fig. 47*, and the spur taken from that; or a separate cable can be run from the ring circuit fuseway in the consumer unit.

Incidentally, if the joint box method is used, the cable should not be cut but simply stripped of its insulation at a suitable point and its conductors should then be laid in the terminals of the joint box.

Although the main purpose of spurs is to supply socket outlets in more remote positions, they are also used to feed a fixed (not portable) appliance such as a wall fire, for example. In these cases, a fused switched connection unit (formerly called a spur box), *fig. 48*, is used instead of a socket outlet at the end of the spur cable. The fused connection unit is

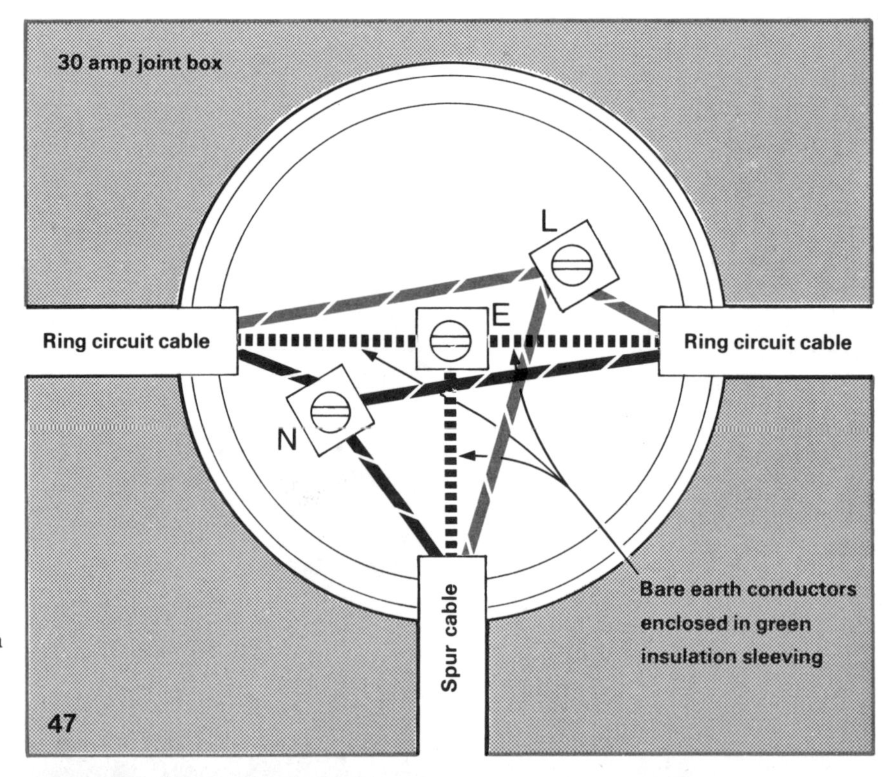

Fig. 47 Spur connected to 30 amp joint box

Fig. 48 Switched fused connection unit

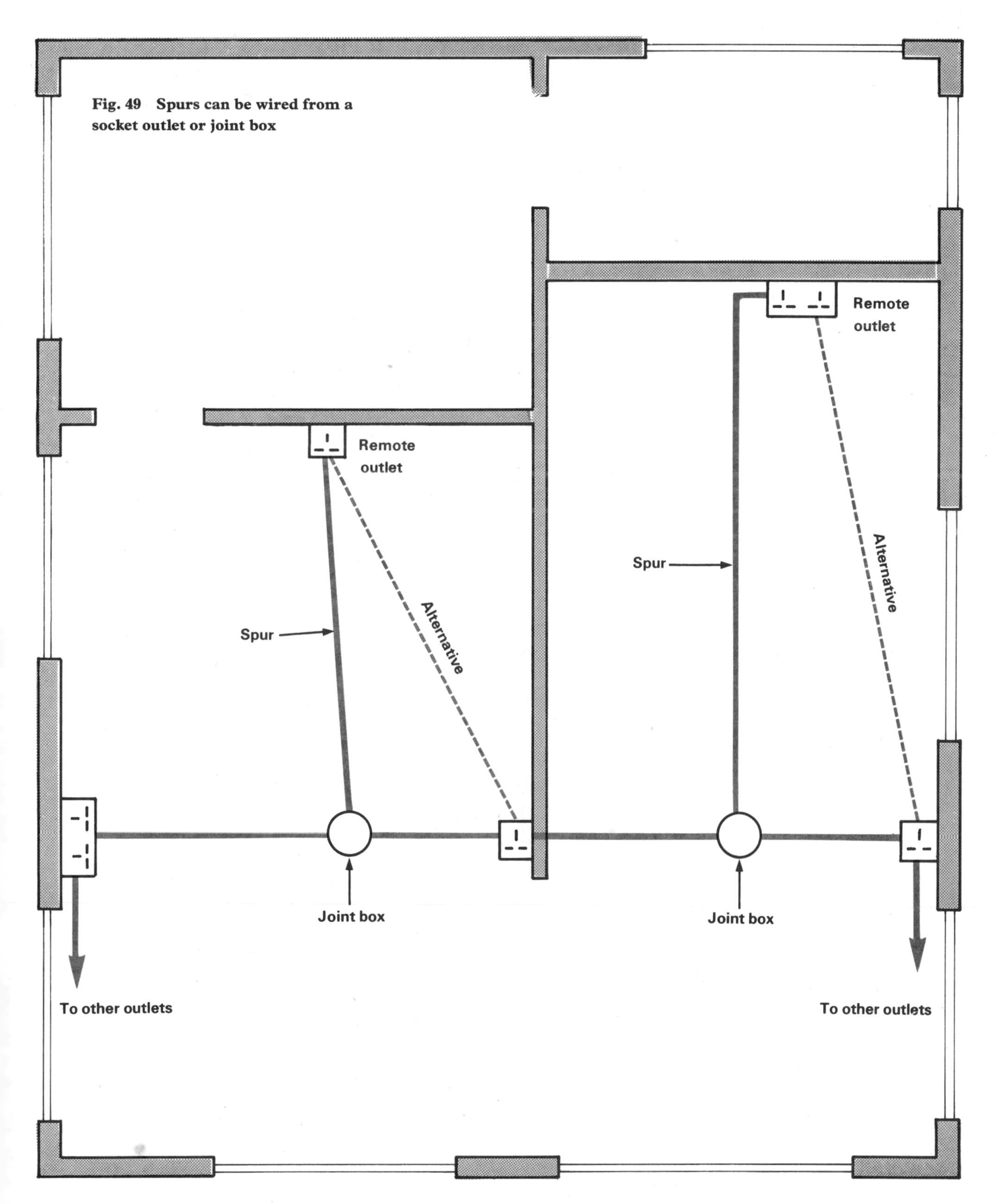

Fig. 49 Spurs can be wired from a socket outlet or joint box

positioned near the fixed appliance and joined to it as shown in *fig. 50*.

It is an advantage if the fused connection unit is a type which has not only a switch to master-control the appliance, but a pilot light also, to indicate whether it is switched on or off.

The fuse in a connection unit is similar to that in a 13 amp fused plug. It can be either a 3 amp or 13 amp type fuse, according to the wattage of the fixed appliance. The fuse is housed in the front of the unit and is held by a small screw. It is essential to fit a fuse of the correct rating.

THE CABLE

The size of the cable now used in a ring circuit is the metric single strand 2·5 mm.². The most common type of cable now used in domestic installations is pvc sheathed twin and earth. Some houses, however, are wired in TRS (tough rubber sheathed) cable of the same size.

Many existing ring circuits, of course, are wired in the old imperial size cable which had seven strands and was known as 7/029 (see separate chapter on cables). The new metric size cable can be used with the old type when extending a ring circuit.

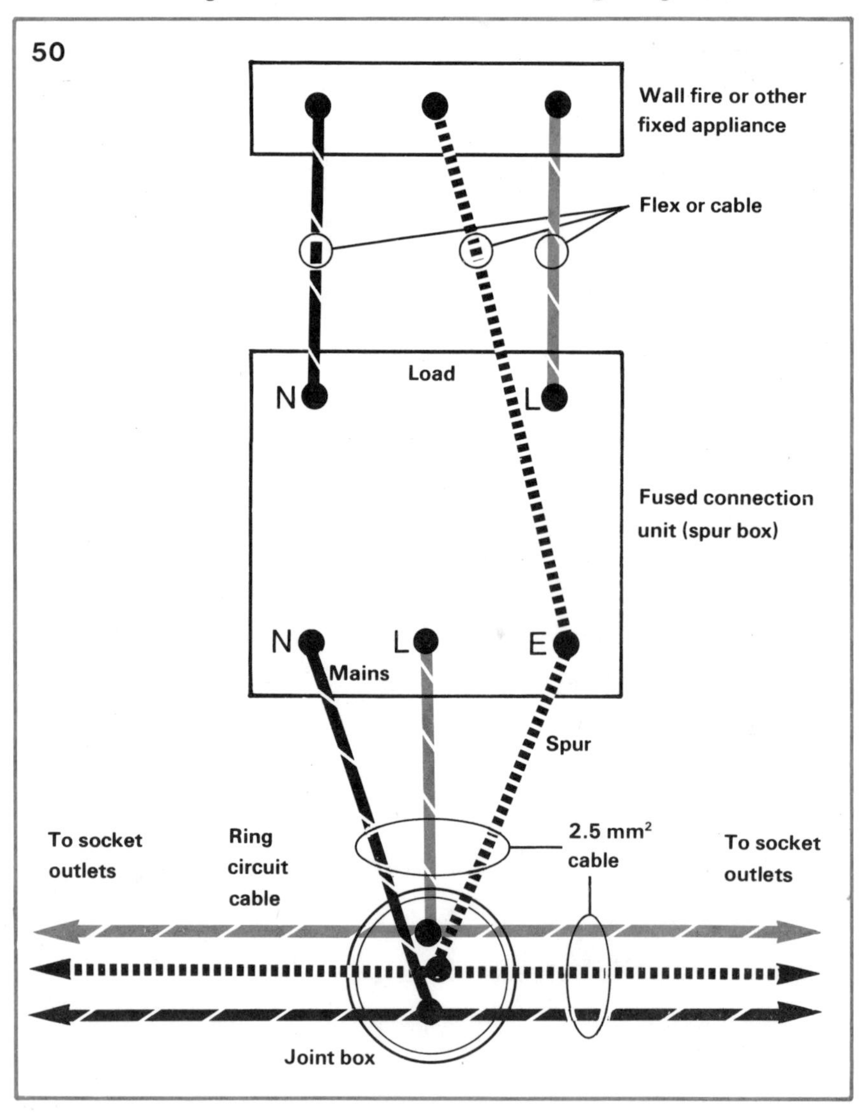

Fig. 50 Fused connection unit wired from joint box

Fig. 51 Ring circuit with three spurs

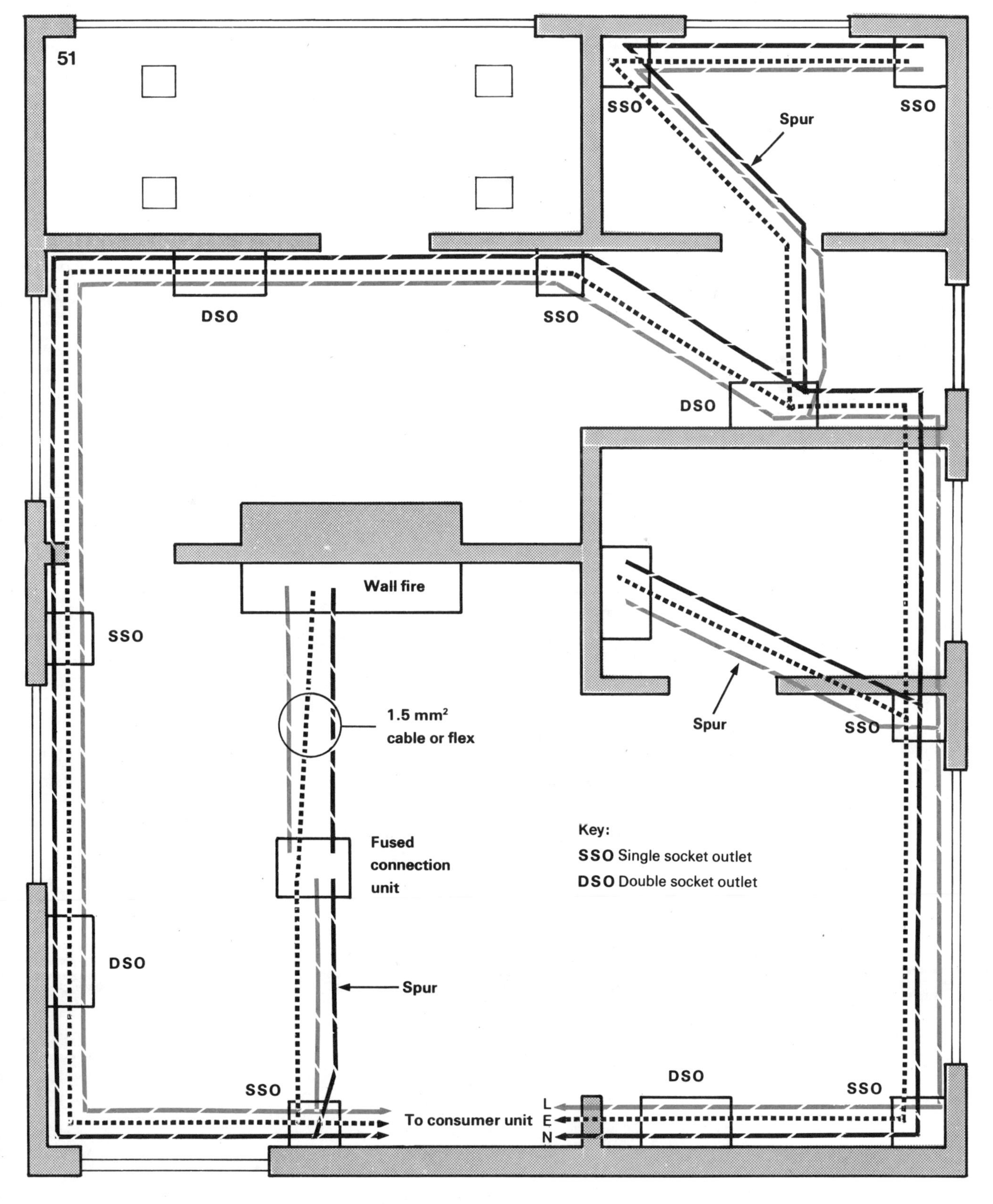
51
SSO
SSO
Spur
DSO
SSO
DSO
Wall fire
SSO
1.5 mm²
cable or flex
Spur
SSO
Fused
connection
unit
Key:
SSO Single socket outlet
DSO Double socket outlet
DSO
Spur
SSO
To consumer unit
L
E
N
DSO
SSO

SOCKET OUTLETS

Some socket outlets used are not fitted with a switch. The tendency now, however, is to install more of the switched type and fewer of the unswitched. Other types used are not only switched, but have a pilot light also, *fig. 52.*

Double (or twin) socket outlets have the obvious advantage of providing two outlets adjacent to each other. This eliminates the need for plug adaptors, *fig. 53*, to a large extent, when a socket is required for more than one appliance.

For example, a standard lamp can occupy one socket and an electric heater the other; or a TV set can be connected to one outlet and a radiogram to the other.

When a house is being rewired, it is wise to have as many double socket outlets installed as possible. The double types are also available with switches and/or pilot lights if they are required.

Your ring circuit, of course, can consist of all single socket outlets, all twin outlets or a mixture of both. Remember, though, that a spur can feed only two single outlets or one double outlet.

ADDING AN OUTLET

One of the biggest faults in some houses (even modern properties) is the inadequate number of socket outlets provided. Adding an extra outlet to a ring circuit is a job a handyman can do. If, though, you have never attempted any wiring work before, make sure you understand what the operation involves.

There are various ways in which an extra outlet can be added. One method is to run a spur from the nearest, or most convenient, existing outlet to the position required. If you decide to do this, first turn off the main switch at the consumer unit and undo the screws on the front plate of the nearest existing socket outlet. Pull the outlet gently away from its wall box and examine the cables to make sure the outlet is not itself a spur.

If there are two red, two black and two earth (green or possibly bare) conductors in the box, the outlet is probably not a spur. I say 'probably' because it could be the first single socket outlet of two on a spur.

The way to check this is to examine also the nearest two socket outlets on either side of it or, better still, trace the route of the cable.

If there is only one set of conductors

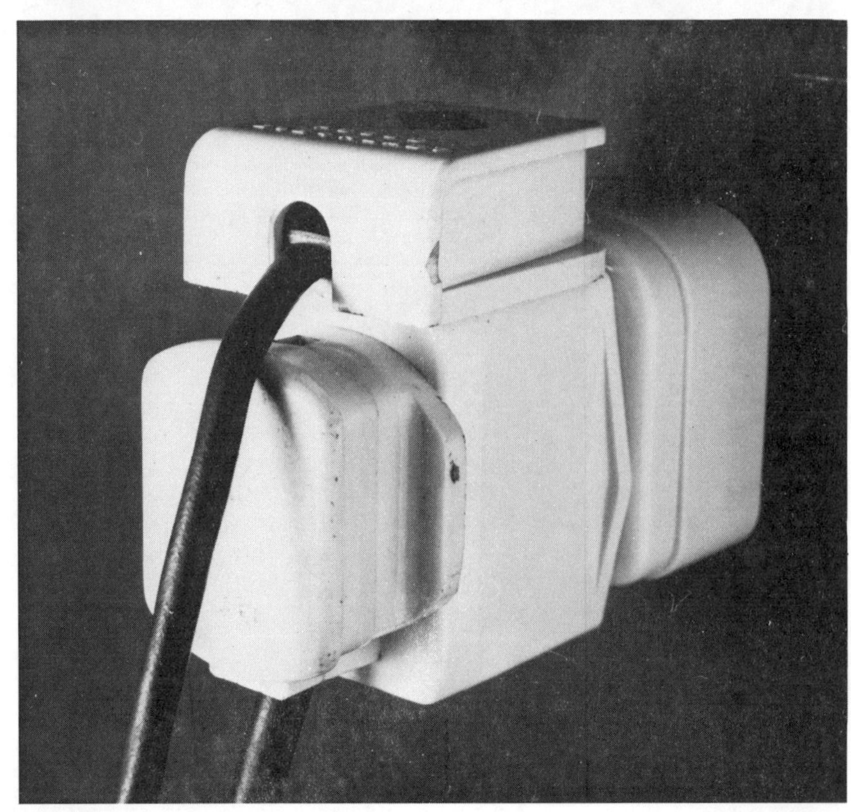

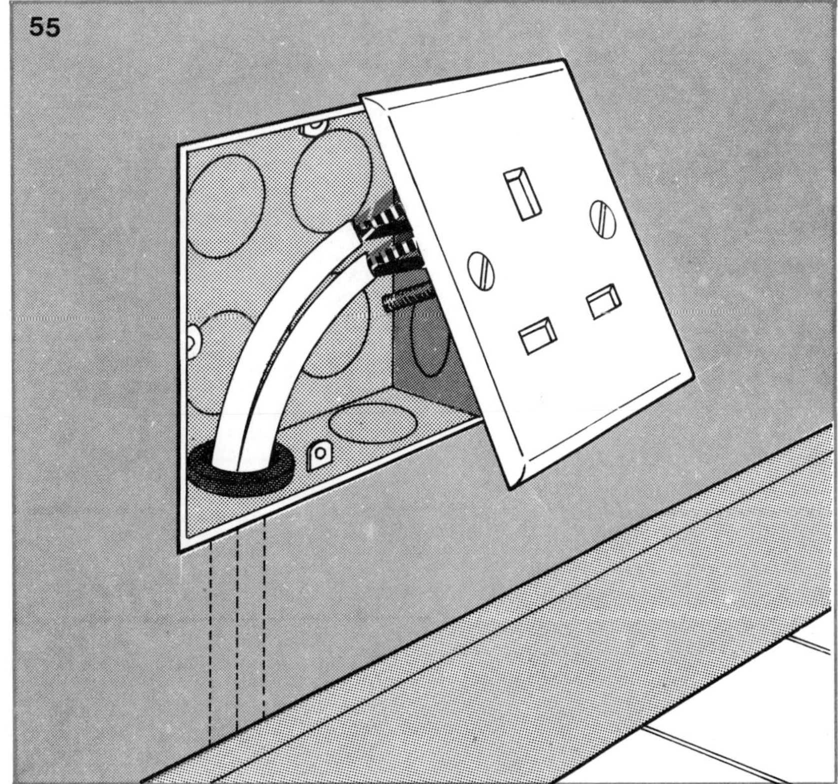

Fig. 52 Twin flush switched socket outlet with pilot light

Fig. 53 Two-way plug adaptor

Fig. 54 Socket outlet on a spur

Fig. 55 Socket outlet on a ring circuit

on the existing outlet, then this will be a spur and should not be used for this purpose.

If the outlet is not a spur, you can now replace the examined outlets and restore the current on the circuit until you are ready to start work.

WHAT YOU NEED

Buy a suitable socket outlet and, if the new outlet is to be a flush type, also buy a matching steel box to accommodate it. If your existing outlets are surface mounted types, you will want to match the new with the old. You will therefore need a matching backplate for the outlet, *fig. 57*, but for the purpose of this book we will assume you are using the flush type.

Also you will require a couple of ¾ in. grommets (plastic or rubber) to fit in the knockout holes of the wall box, *fig. 58*, to protect the cable; a suitable length of 2·5 $mm.^2$ pvc sheathed twin and earth cable; and a short length of green sleeve insulation. This insulation sleeving is to slip over the bare end of the earth conductor in the steel box to prevent the bare wire fouling another terminal if it should at some time work loose.

As it will be necessary to channel out a groove for the cable on the plastered wall from the old outlet to the new, the best time to do this job is before redecorating. Alternatively, it may be possible to run the new cable under the floorboards or behind the skirting board, *fig. 60*. Much will, of course, depend on circumstances. If you plan to redecorate in the near future, the cable can be fixed temporarily on the surface of the wall.

FIXING THE WALL BOX

Mark out the position of the new socket outlet by holding the steel box to the wall and make a pencil mark around it. Score the lines with a sharp handyman's knife to break the plaster surface. Then with an old but sharp chisel, tap out the plaster back to the brickwork. Avoid damaging the

58

Fig. 56 Backplate for single socket outlet for surface mounting

Fig. 57 Wall box for single socket outlet

Fig. 58 Grommets for a knock-out box

Fig. 59 Step-by-step stages for the installation of a recessed socket outlet

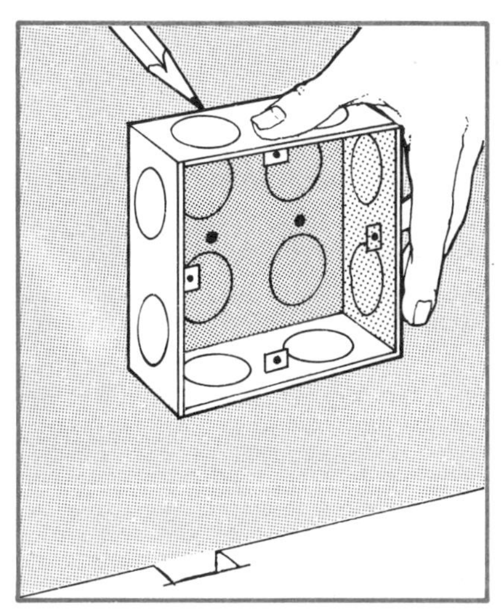

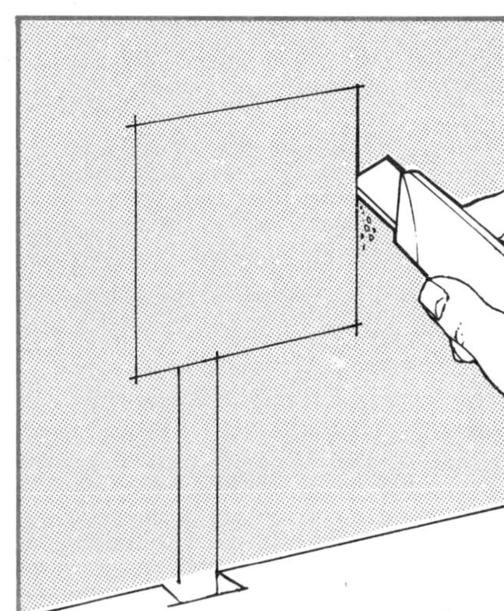

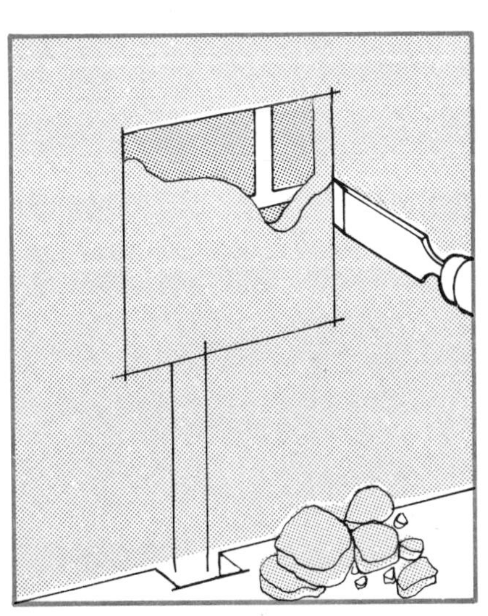

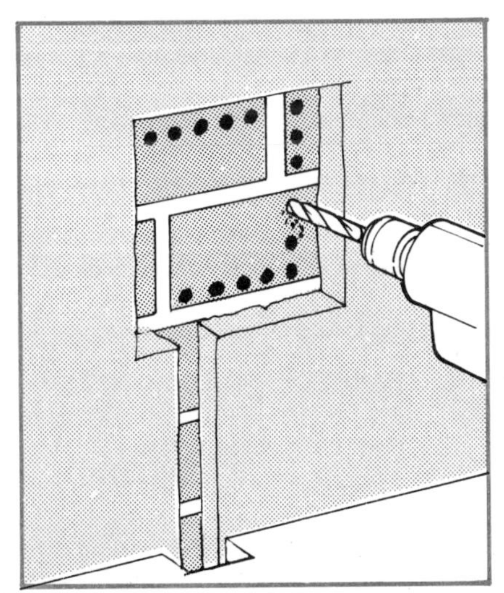

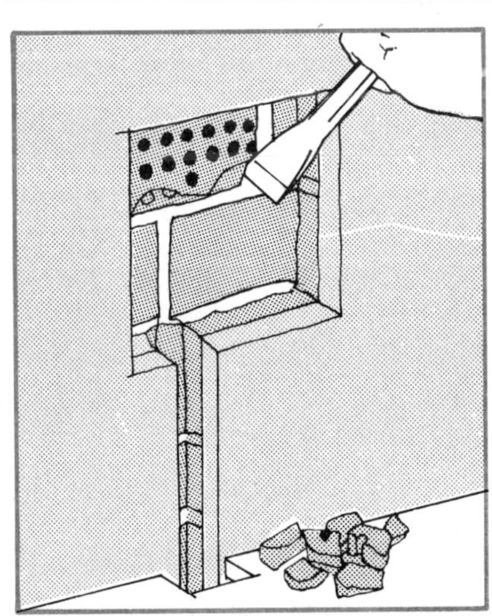

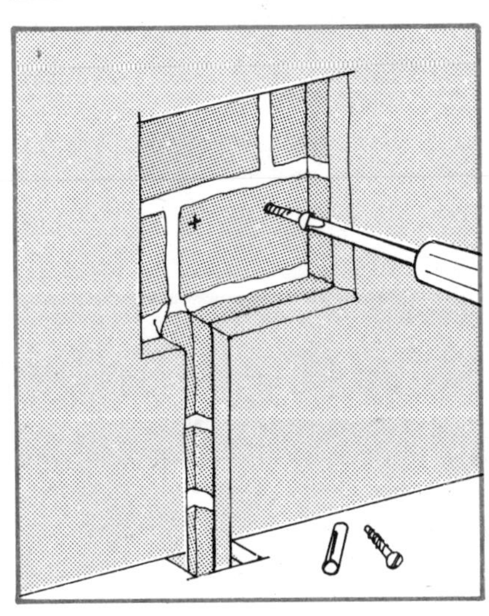

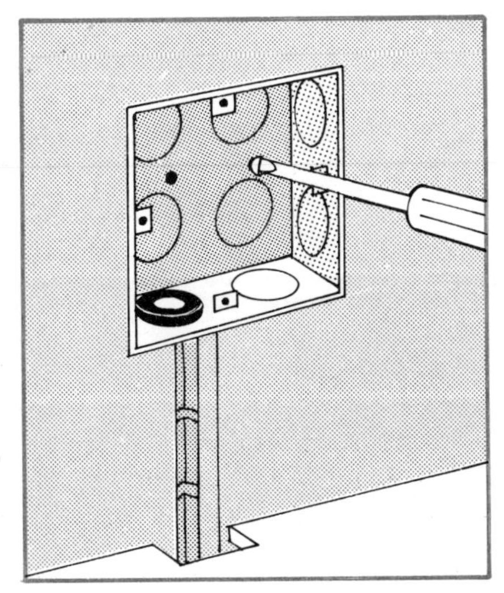

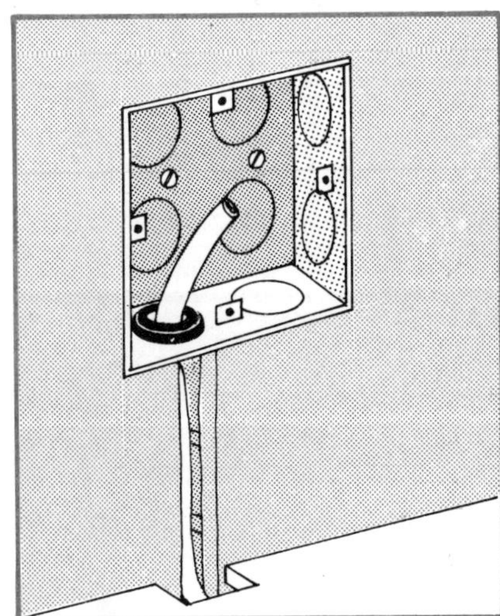

surrounding plaster as much as possible.

Another method is to drill a series of small holes within the marked off area, to produce a honeycomb effect, and remove surplus plaster with the chisel.

You will require a hole about $1\frac{3}{8}$ in. deep, so a certain amount of brickwork will also need to be removed. It helps to drill a series of holes with a masonry drill first and remove the surplus brickwork with a cold chisel. Make sure the hole is deep enough by trying the box in it at intervals.

GET IT LEVEL

When the box fits snugly, hold it firmly in place, make certain it is level, and make a mark on the wall for the fixing screws at the back of the box using a sharp tool through the fixing holes of the box. Drill and plug the holes for the screws and drive them partly in. Check again that the box is level and then withdraw the screws.

The box will have a number of knock-out holes, some of which may be marked M (for metric size). Remove a suitably positioned knockout using an old nail punch and hammer, or just the hammer. The knockout can be finally pushed out, *fig. 61*.

Insert a grommet in the hole knocked out and thread about 5 in. of cable (no more) into the box, *fig. 62*.

Fig. 60 Adding an outlet; alternative cable routes

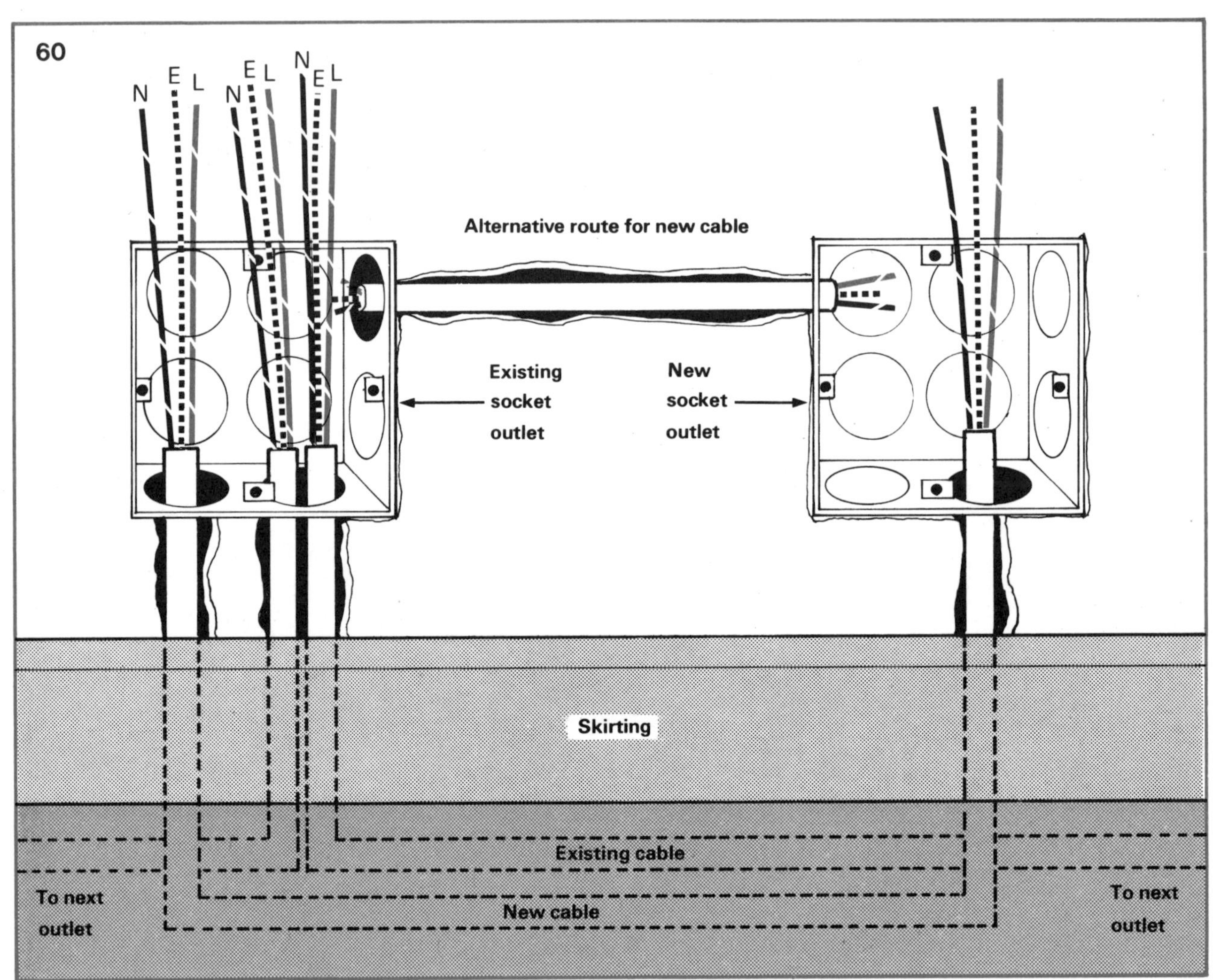

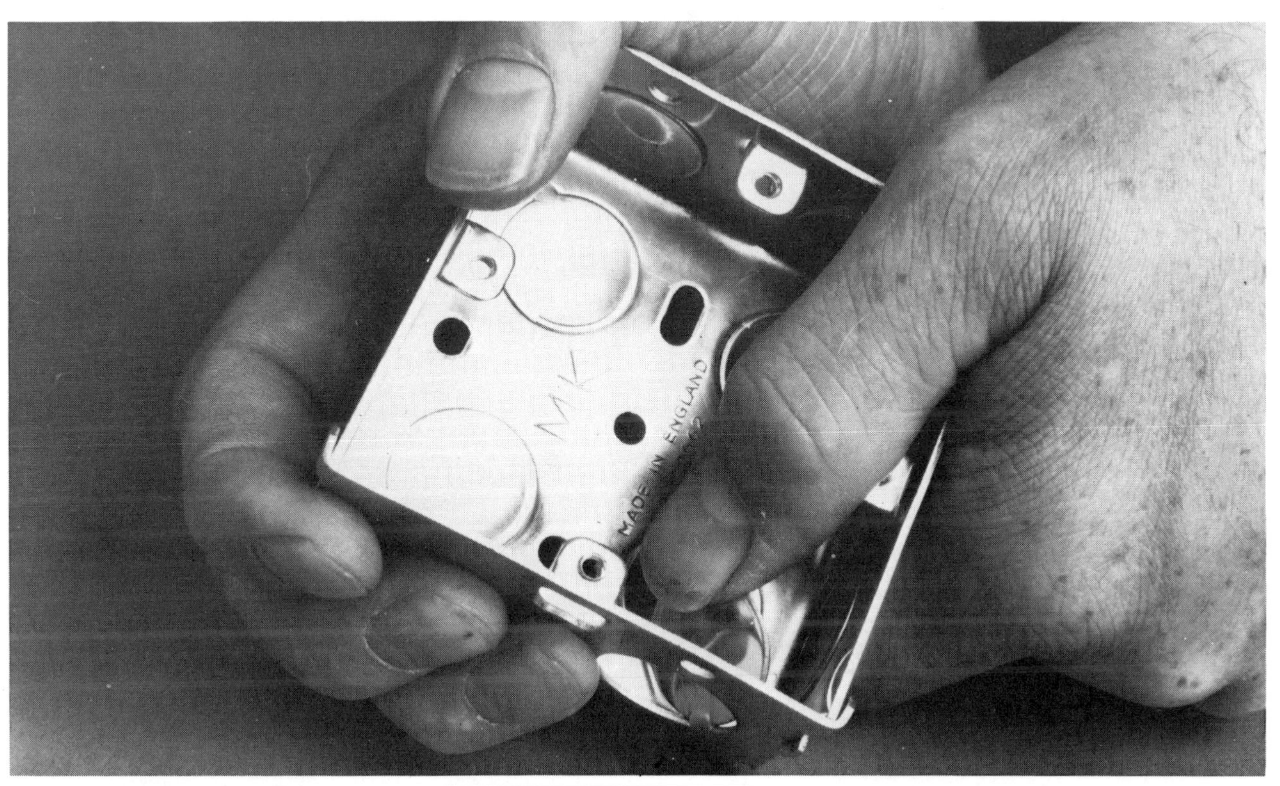

Fig. 61 Pushing out a knockout

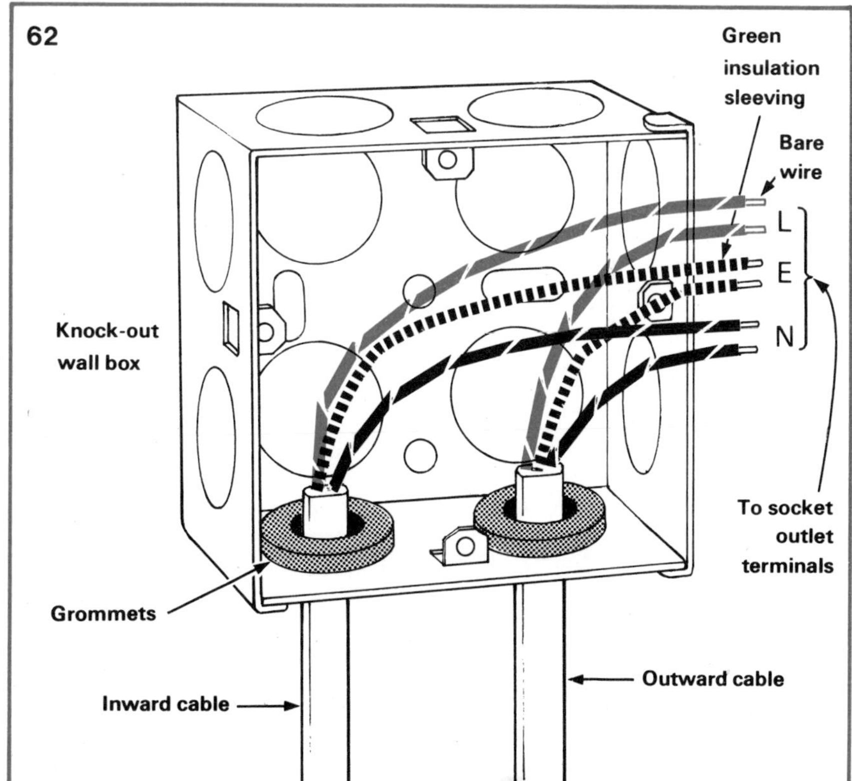

Fig. 62 Ring circuit cables ready to connect to an outlet

CABLE CHANNEL

If the new cable is to be buried under the plaster, mark out its route by drawing two parallel lines the width of the cable apart. Score the lines with a sharp knife to break the plaster.

The plaster can then be removed in the same way as for the outlet's wall box, starting from the new outlet position.

Another way to do this is to drill a series of holes along the route and chase out the plaster with a simple little tool called a Cintride router, *fig. 63*. This can be fitted into an electric drill run at a slow speed.

Unless the plaster is very thin, it should not be necessary to remove any brickwork, but if it is, this can be done by drilling several holes with a masonry drill and removing the surplus brick with a cold chisel. Make sure that the channel is deep enough all along its route, for nothing looks worse than a cable bulging even slightly from a wall.

Stop well short of the existing socket outlet, *turn off the current*, and then complete the channel.

Turning off the current is a safety precaution which should be taken from the outset if you are not sure where existing cables are situated.

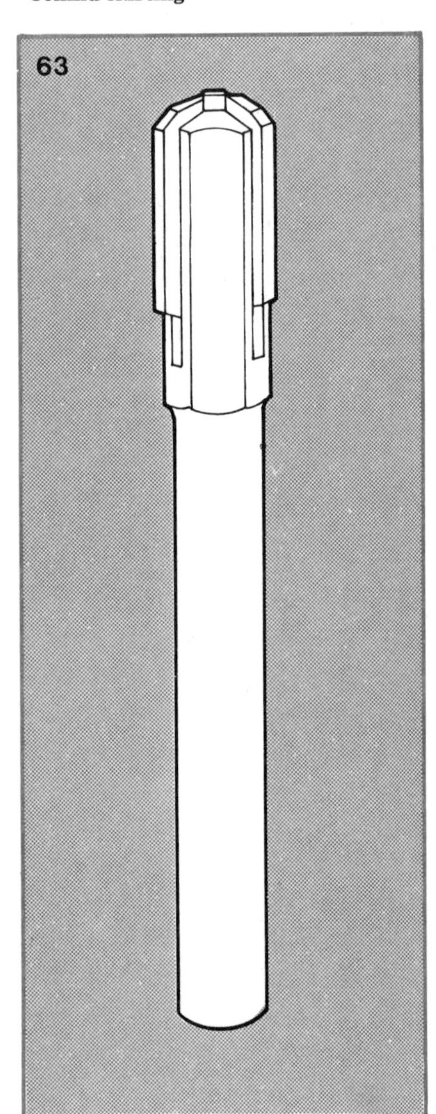

Fig. 63 A Cintride router for chasing out wall plaster

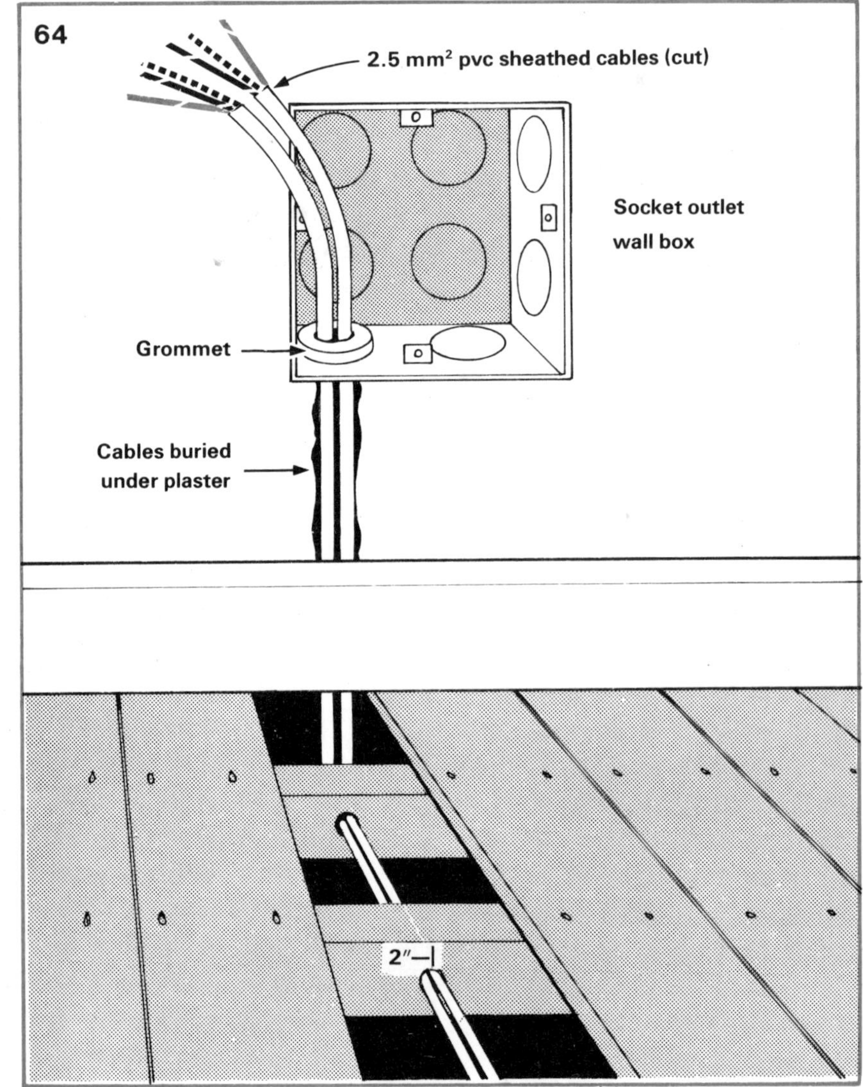

Fig. 64 Ring circuit cables rising behind skirting

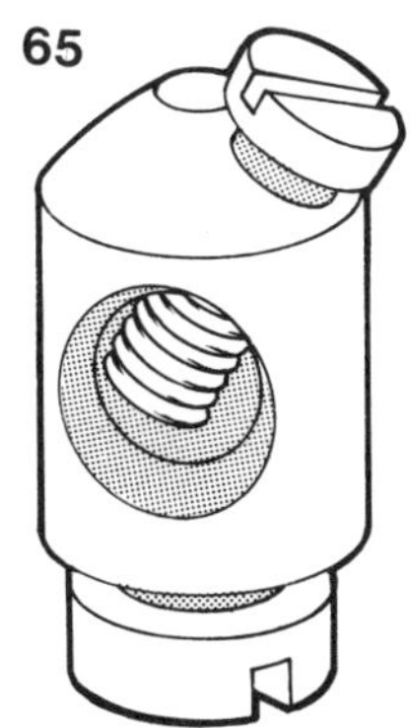

Fig. 65 Earthing terminal

EXISTING CABLES

Now release the fixing screws of the existing socket outlet, pull it gently from its box and unscrew the terminals to free the conductors.

You will probably find that the cable comes up from below into the steel box, *fig. 64*. As I said earlier, there should be two black, two red and two green conductors, twisted together in pairs, in the box. These are the neutral (black); live (red); and earth (green). Leave these in their twisted state. They will be the old imperial size, seven strand conductors. If, however, yours is a modern installation, you will find the conductors are single strand (metric size).

Your new cable will, of course, also be single strand and the conductors will be difficult to twist around old seven-strand wires. Don't try it! Let the end of each new conductor enter the terminal *beside* the twisted wires, but do make sure that the terminal screw bites firmly on to all the conductor ends when you wire up.

The next step is to remove the existing wall box to knock a hole in it (a knockout) for the extra cable. (It is sometimes possible to remove a knockout from a box while it is still in position in the wall, but it is not very easy. Also, it usually involves damaging the surrounding plaster).

The old box should be fixed with screws at the back, but it may be simply cemented in the wall.

You will probably have to chip away a fair bit of plaster to get the old box out and it may become distorted in the process. Have a spare one handy in case you need it.

Fit a grommet in the knockout hole.

If the house is wired in metal conduit, there may not be an earth wire. Metal conduit can act as the earthing conductor. However, your new socket outlet must be earthed. So fit an earthing terminal, *fig. 65*, to your old wall box if one is not already fitted.

At this stage your new cable can be finally cut to size. Fit it into the channel but do not plaster over it yet. The cable may have to be temporarily wedged in place with thin pieces of wood. This size of cable is tough stuff with a mind of its own and will probably jump out of the channel if not wedged!

Thread about 5 in. of cable into a knocked out hole in the existing box and remove about 4 in. of outer sheath. You may find it easier to do this before finally threading the cable into the box.

JOINING UP

Strip about 1 in. of insulation material from the red and black conductors. The earth conductor will be bare. Slip a short length of green insulation sleeving over it, leaving about 1 in. of its end bare.

Arrange your new conductors neatly with the old ones (black next to black; red with red; green with green) so that they can conveniently enter the terminals.

Take the two (twisted) old red wires and single (new) red wire to the terminal marked L for live. Connect the black wires to terminal N for neutral; and the greens to terminal E for earth. Tighten the terminals; make sure the ends of the conductors are firmly secured by the terminal screws. Then replace the socket outlet.

If your house is wired in conduit, run a short length of single core bare wire covered with green sleeving from the earth terminal you fitted on the old wall box to the earth terminal on the new socket outlet. This will ensure good earth continuity to the new outlet.

The new socket outlet is wired up in the same way except, of course, that there is only one set of conductors.

When both socket outlets are wired up and the cable temporarily in place, turn the current on and try both outlets with an appliance you know is working.

If all is well, the channel can then be made good with a plaster filler, but to be on the safe side, turn the current off again before doing so. If the cable is reluctant to stay in position in its channel, try fixing it with a few spots of contact adhesive.

USING A JOINT BOX

Another method of increasing the number of socket outlets is to break into the wiring of the ring circuit at a suitable point and add a spur cable. Again, make sure you do not break into an existing spur for this purpose.

To add the spur cable, you will need a 30 amp joint box, *fig. 66*, which can be fitted on a short length of timber fixed between joists, or to other suitable timber fixture within reach of the cable.

The terminal box has three terminals – for live, neutral and earth conductors – and knockouts to permit cable entry. The middle terminal is used for the earth conductors. It does not matter which of the other terminals is used for live or neutral so long as you are consistent!

Remove suitable knockouts for the ring circuit conductors – one for their entry; another for their exit. Remove one other knockout for the spur cable. Screw the box in position.

Prepare one end of your new cable as for entry into a socket outlet. Don't cut the ring circuit cable if you can avoid doing so. Simply strip the sheath and the insulation at a suitable point so that there is enough bare wire of

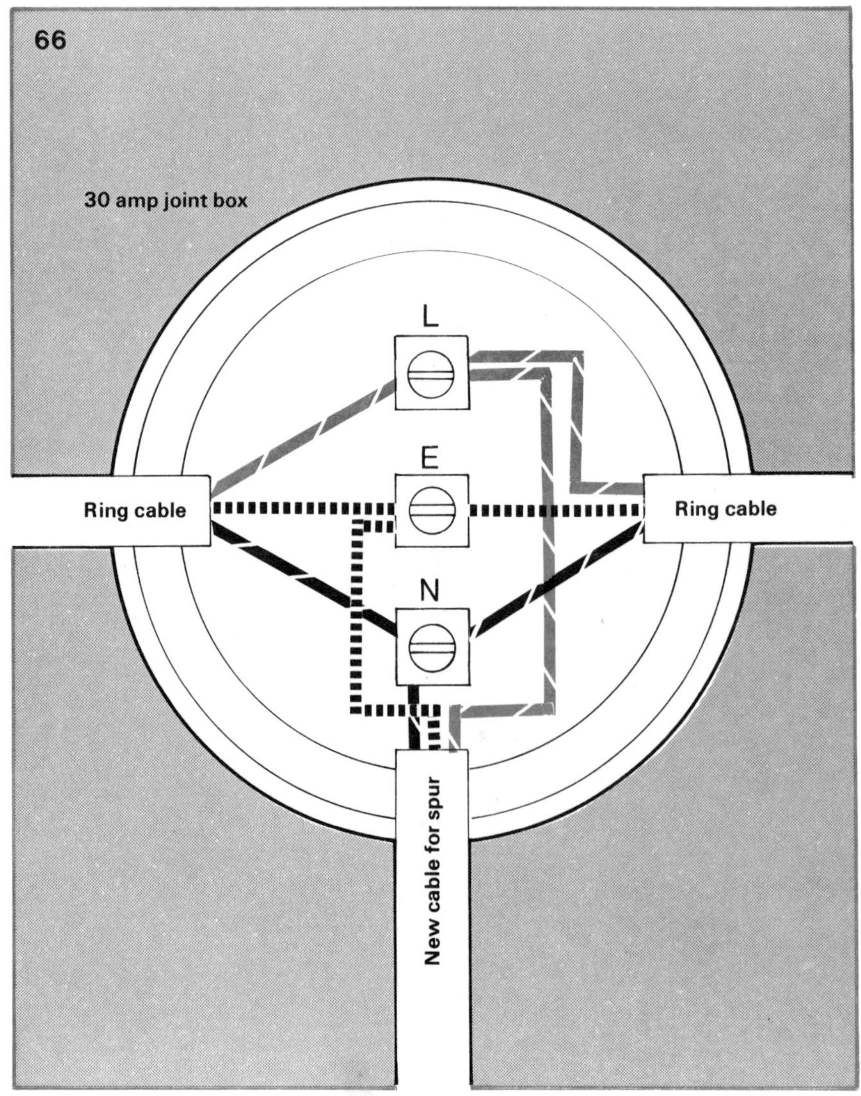

Fig. 66 Joint box used to supply a ring circuit spur

each conductor to lay in the terminals of the joint box under the terminal screws.

Tighten the terminal screws and replace the joint box cover.

The other end of the new cable is then taken to the new socket outlet which is fixed as already described.

EXTENDING THE RING

If you want to add more than one socket outlet to a ring circuit, this can be done by opening up the ring at a suitable point and extending it.

This is subject to my earlier remarks regarding the size of the floor area; i.e., you need more than one ring circuit if that area is greater than 1,080 sq. ft.

To add the outlets, all you do is disconnect either the inward or the outward cable at a socket outlet. Take this cable to the first of the new socket outlets A, via a joint box if too short.

The first outlet, in turn, is linked to the next outlet and so on likewise until the last of the new socket outlets is reached. This outlet is then linked to the original socket outlet where the ring was broken into. Thus, the continuity of the ring is maintained. *Figure 67* is simply an example.

Fig. 67 Three outlets added to a ring circuit

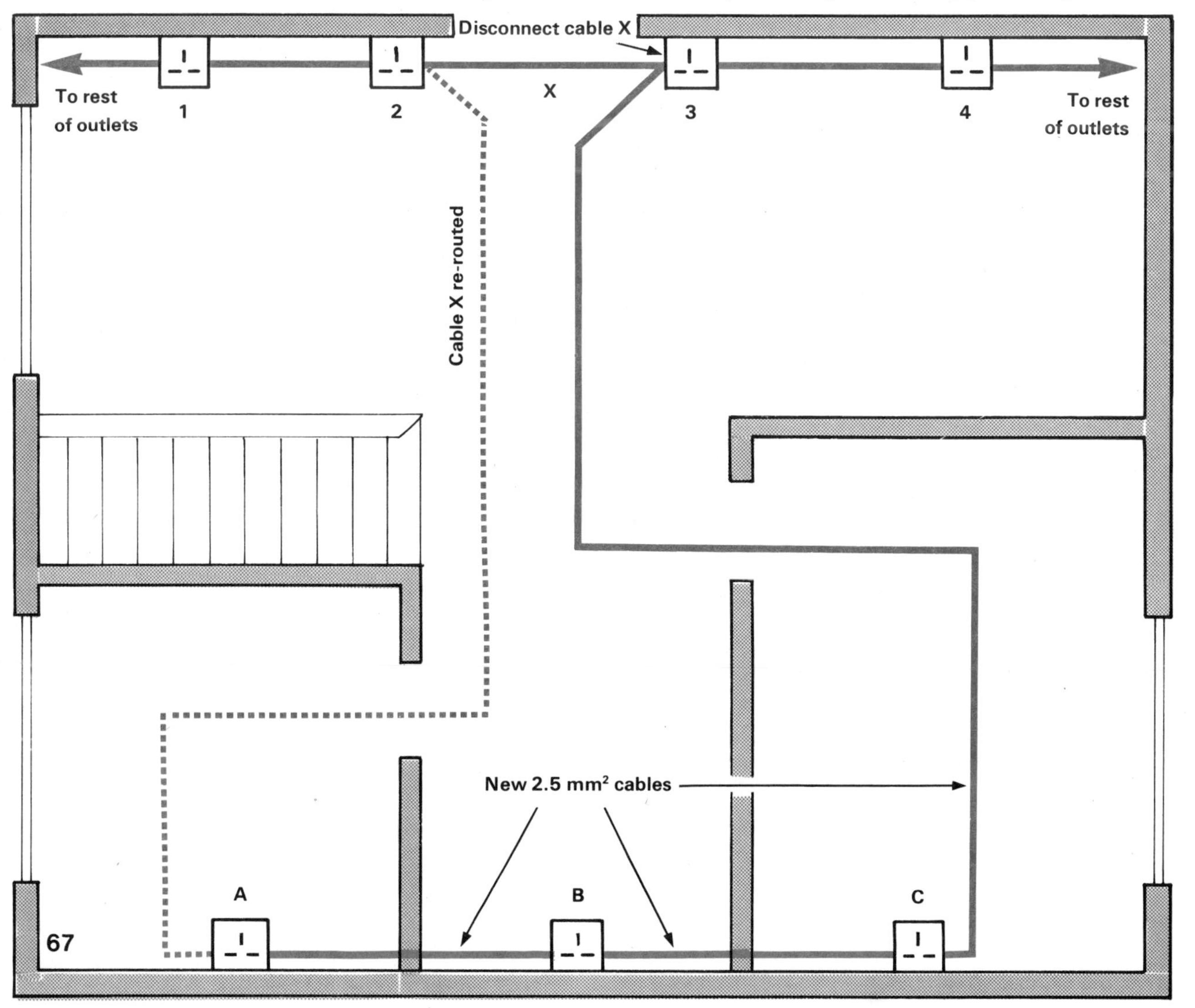

SOLID FLOORS

Many modern houses have solid ground floors which can present problems when running cables after the house is built. Obviously, they cannot be buried under a concrete floor at a later stage.

One way out of the difficulty is to drop the ends of the cables down through the ceiling to the floor below immediately above the position planned for the new socket outlet. The cables can then either be buried under the wall plaster as described earlier, or can be run on the surface of the wall.

This method will, of course, take a lot of cable as the cables will, in all probability, first have to be taken from the consumer unit on the ground floor, up to the first floor (under the floorboards) and down again.

A possible alternative is to take out the skirting board and replace it with special trunking designed to house cables. This is available in metal and plastic.

The solid floor problem can be overcome if you want to add only one or two outlets by using the method of burying the cables in the wall plaster and looping out of existing outlets as described.

FIXING WALL BOXES

Instead of fixing a metal type flush wall box to house an outlet or other accessory, you may be tempted to think of an easier way of doing the job than cutting out a hole in the plaster and brickwork to accommodate it.

In this case, the best plan is to use a surface-type box with its appropriate surface type accessories. Whatever you decide to do, don't be tempted to fix a wall box in the skirting board. This is against the regulations and can be very dangerous.

Equally, if not more, dangerous is the practice of fitting a socket outlet to a skirting or to a wall without using any sort of box. This should never be done.

The danger is that, although the brickwork may be non-combustible, there is always the possibility of condensation running down the wall or damp creeping up it and finding its way to the terminals of the outlet.

The regulations lay down that cores of sheathed cable from which the sheath has been removed should be enclosed in non-combustible material.

The boxes sold to accommodate either flush or surface type accessories comply with these requirements.

Lighting Circuits

A lighting wiring system is usually regarded as complicated. In fact, basically, this is not so; it is the number of wires which are gathered together at each point which make it seem so.

Like a ring circuit, a lighting circuit starts from the consumer unit. It runs to all the lighting points on that circuit and finishes at the last one. Unlike the ring circuit cable, the lighting cable does *not* return to the consumer unit, *fig. 69*.

Each cable contains three conductors – live, neutral and earth. Old installations have no earth wire, but this is now compulsory in new wiring. The regulation, however, is not retrospective; it applies only to new wiring.

The cable used in a lighting circuit can be either 1·0 mm.² or 1·5 mm.² flat twin core and earth pvc sheathed.

TWO SYSTEMS

There are two basic ways of wiring a domestic lighting circuit – the looping in system or the joint box system. There are some installations which use a bit of each.

Generally speaking, the looping in method is the more popular. The other method is for surface wiring where the looping system is not suitable. Wall lights are one example. They have no facilities for looping in and, of course, no ceiling roses.

LOOPING IN

With the looping in method the cable runs from the consumer unit to each ceiling rose in turn. The cable conductors loop into and out of each rose terminal until the last rose is reached. This is where the cable terminates.

Fig. 68 Live and neutral connections at a rose

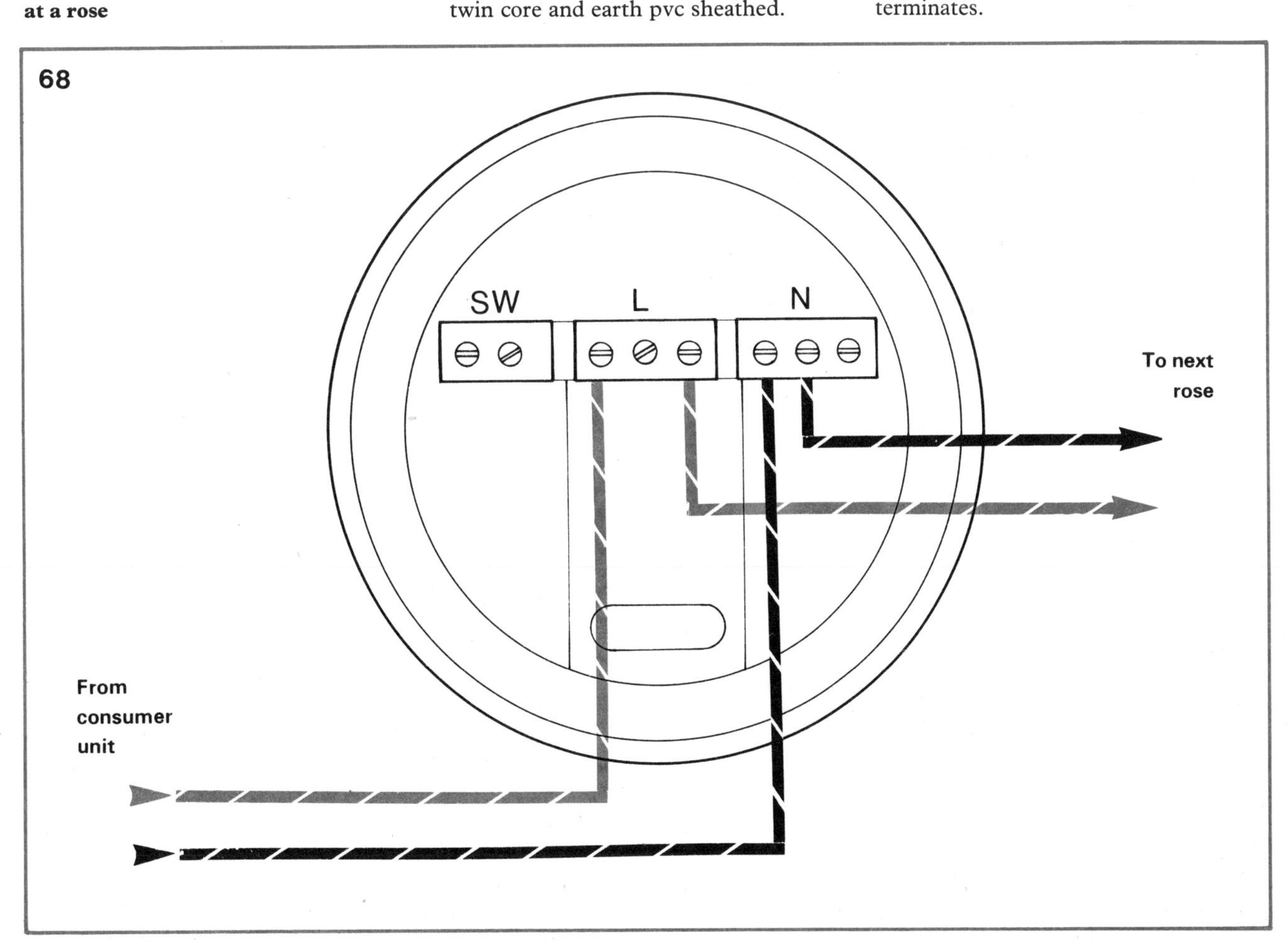

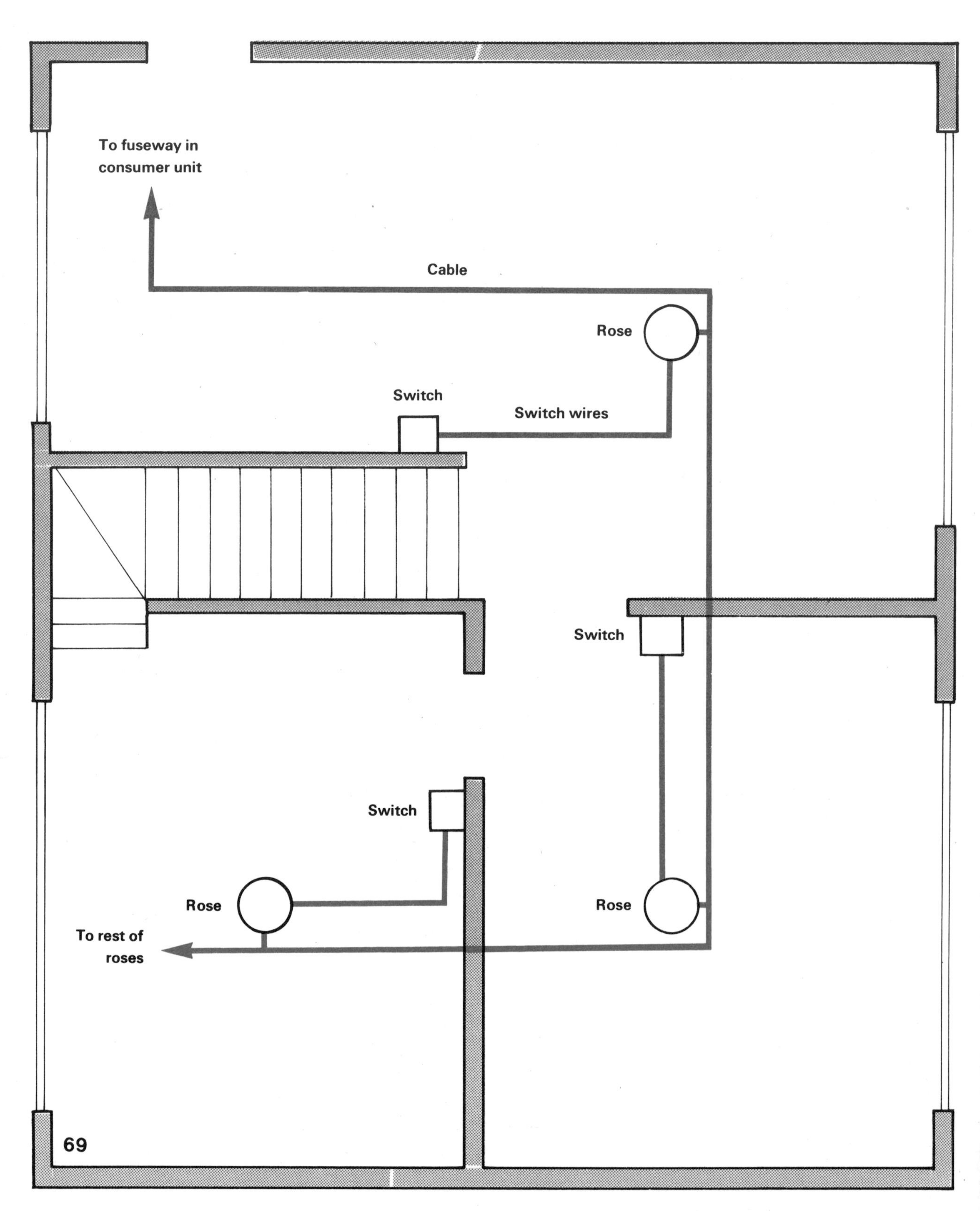
To fuseway in
consumer unit
Cable
Rose
Switch
Switch wires
Switch
Switch
Rose
Rose
To rest of
roses
69

Fig. 69 Basic lighting circuit

Fig. 70 Basic wiring of rose (earth wire omitted)

(It is partly the looping in and out of these conductors which gives the impression of confusion at a ceiling rose.)

Also, a length of cable runs from each rose down to the switch on the wall. *Figure 68* illustrates the basic principle. Switch and earth wires have been omitted to make it clearer. *Figure 70* shows the basic wiring details at a ceiling rose.

In most existing installations (apart from brand new property) the wiring will appear as shown – i.e., no earth wire. The earth connections in *fig. 70* have been omitted purely for clarity's sake.

Although the wire from the SW (switch wire) terminal on the rose to the switch is black it is, in fact, the switch return wire and is therefore on the *live* side of the circuit. It is common practice to use the black conductor of a length of twin pvc cable for this purpose.

Therefore, it is never safe to assume that a black wire in a lighting circuit is not live!

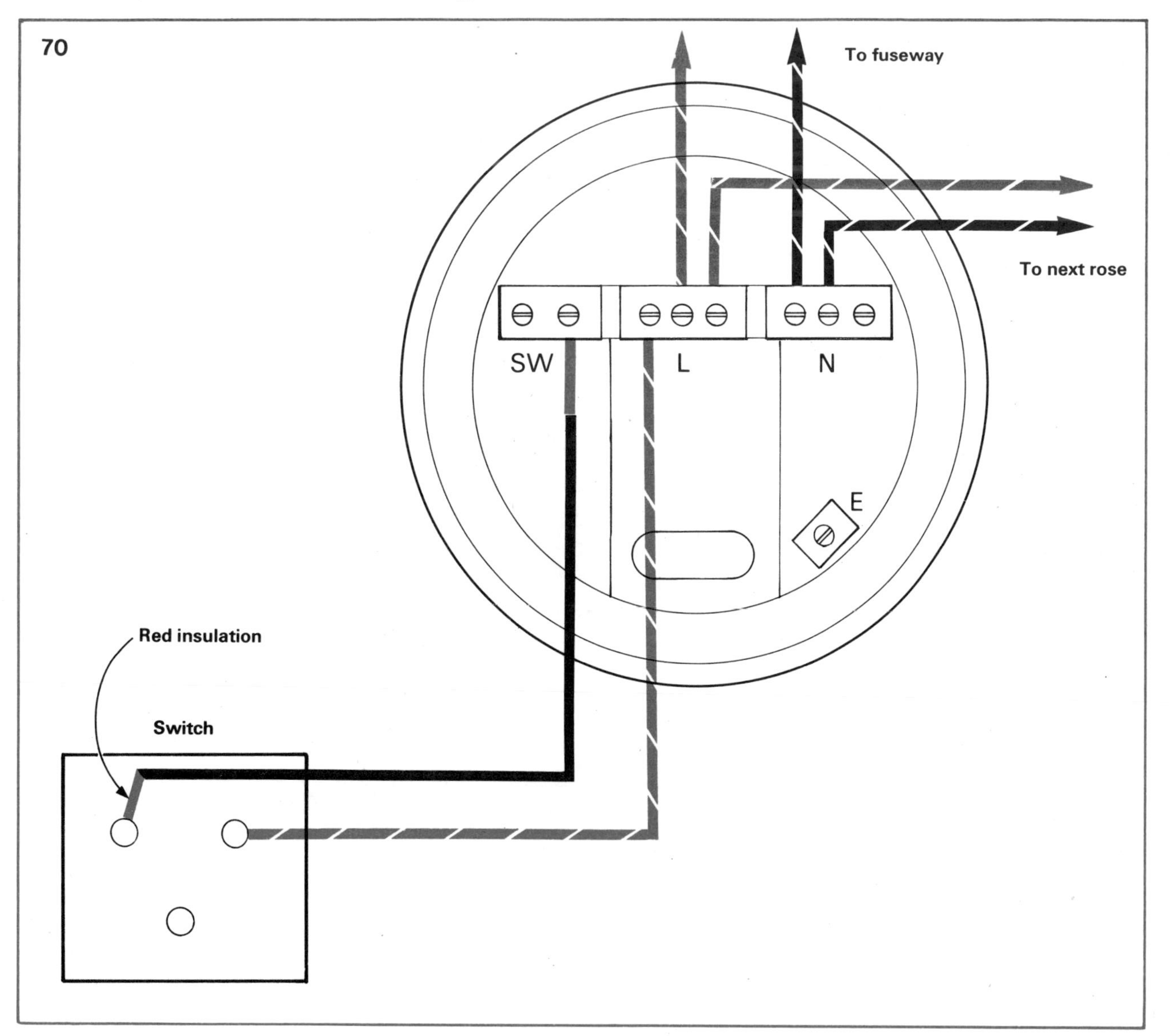

Ideally, the end of the black switch wire should be enclosed in a short length of insulation sleeving so as to identify it as the switch wire. This, unfortunately, is not always done as it should be.

Figure 71 shows the full wiring at a loop-in ceiling rose and illustrates the remarks made about the switch return wire.

Fig. 71 Elaboration of fig. 70

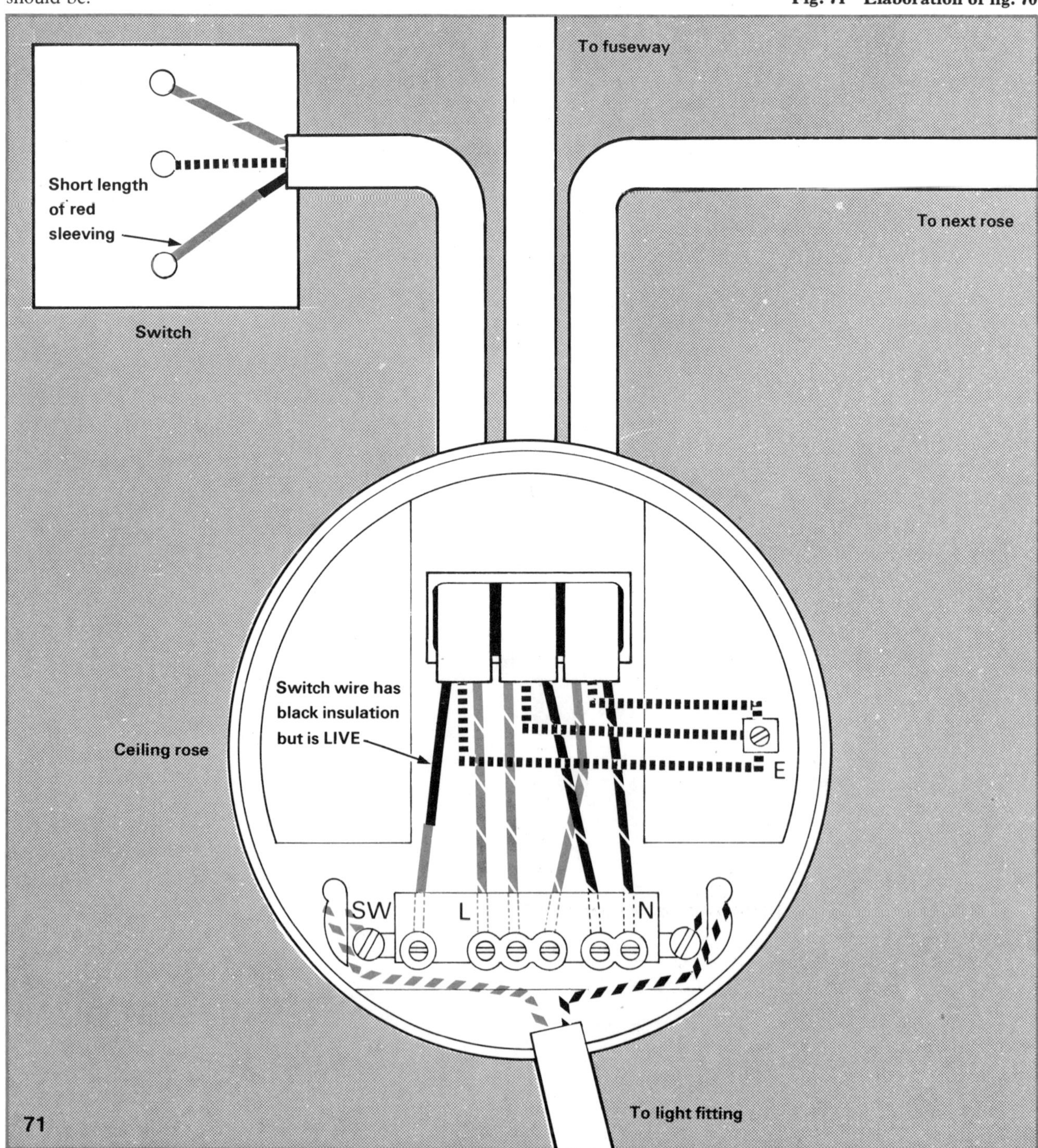

JOINT BOX METHOD

This method resembles the looping in system in some respects, but the feed cables from the fuseway in the consumer unit are looped in and out of joint boxes instead of ceiling roses, *fig. 72*

If your house is wired in the joint box system, you will find a number of small plastic joint boxes above the

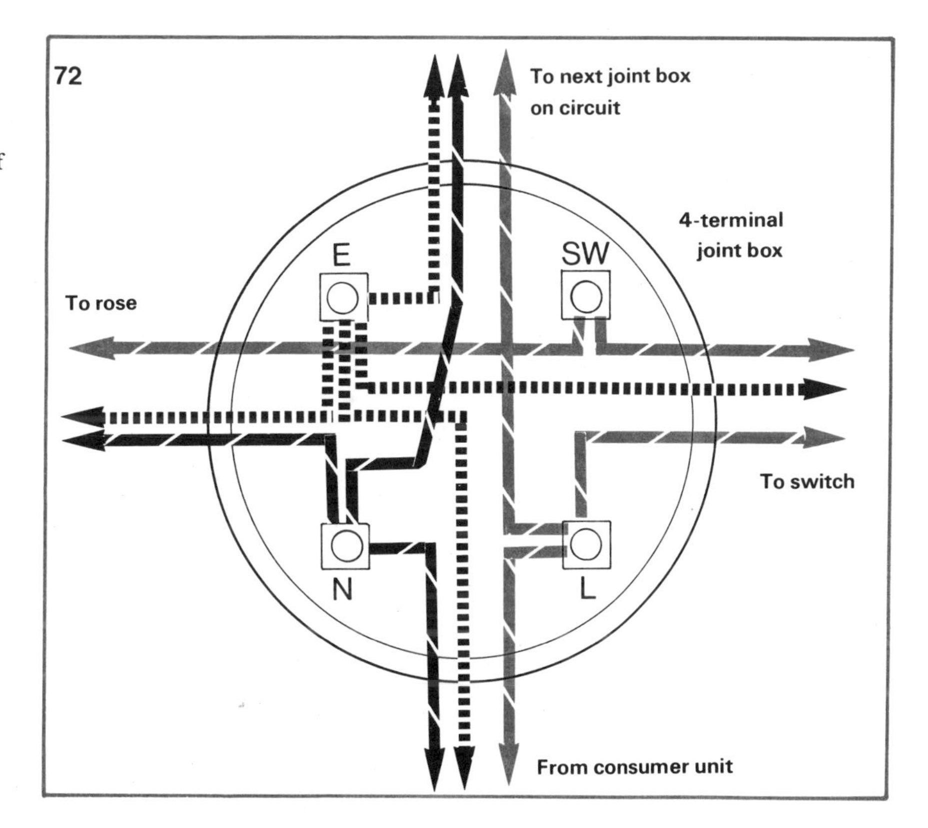

Fig. 72 Joint box connections in a lighting circuit

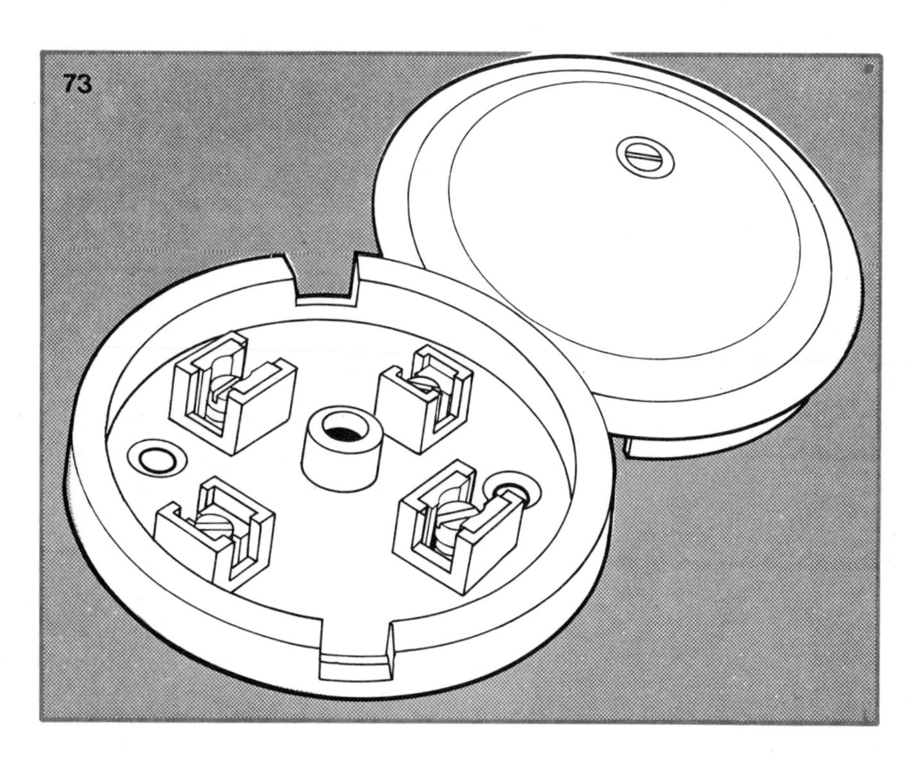

Fig. 73 Joint box for lighting circuits

spots where the cables run down the wall under the plaster to the switches: that is, between joists in the roof space and under the floorboards.

The joint boxes normally have four terminals, *fig. 73.* In an old wiring system, where there is no earth wire, the box may supply two lights, the fourth terminal sometimes being utilised for the switch wire of another light situated nearby.

Some old-type ceiling roses used with the joint box method have only two terminals. This is because no looping into a rose is necessary, each light having its own joint box.

ADDING A LIGHT

Before contemplating the addition of a lighting point to a circuit, it is essential to examine the wiring system thoroughly. If it is a jumble of wires, don't attempt any alterations unless you are absolutely sure what you are doing.

Electrical shops are only too familiar with customers who start a rewiring job but are unable to finish it because they 'don't know where the wires go.' And how can the poor shopkeeper help when he cannot possibly guess which wiring system has been used?

A quick way to establish which method of wiring has been used is to examine a ceiling rose. Turn off the current at the main switch in the consumer unit and unscrew the rose cover.

If the rose has only two terminals, yours is the joint box method and you can confirm this by locating the various joint boxes as described. Three terminals or more indicate that the looping in system has been used.

CONNECTIONS

To add a lighting point to the looping-in system, a twin-core and earth pvc sheathed cable should be run from the nearest convenient existing ceiling rose to the new point, *fig. 74.*

Connect the red conductor to the live (L) terminal and the black conductor to the neutral (N) terminal of the existing rose. If this rose has only three terminals, you will have to terminate the earth conductor in the backplate or pattress and wrap insulating tape around its end to prevent it fouling a terminal. Label it Earth Wire.

For the new rose, buy a loop-in type with an earthing terminal. Connect the other end of the earth conductor to the terminal, but slip a short length of green insulating sleeving over its exposed portion. Then take the red and black conductors to the L and N terminals respectively.

Fig. 74 Wiring up a new rose

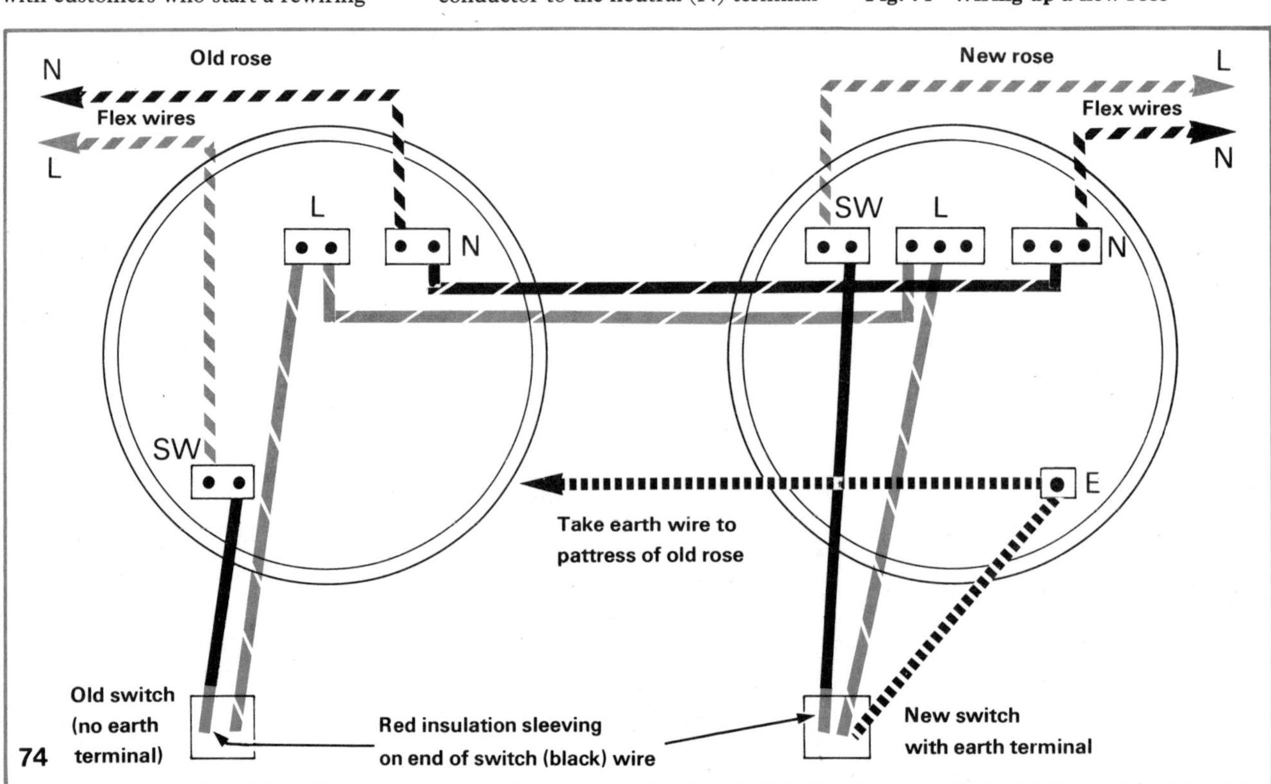

Fig. 75 Modern flush switch

Fig. 76 Plaster depth wall box

SWITCH CONNECTIONS

A length of similar cable is needed to run from the new rose to its switch. The red conductor goes to the L terminal of the rose and the black conductor (the switch wire) to the unoccupied terminal (SW).

The other ends of these conductors go to the switch. Slip a short length of red sleeving insulation over the end of the black wire in the switch box to identify it as the switch wire, or simply label it Switch Wire.

FITTING THE FLEX

That completes the wiring of the cables. Now you need to fit the flex wires. The best type of flex to buy for this purpose is heat resisting circular flex, size 0·75mm.^{2}.

Connect the flex connectors to the rose terminals N and SW, *fig. 71*. Choose a heat resisting type of lampholder which is suitable for the circular flex.

The type of switch used will obviously be chosen so as not to clash with the appearance of the others on the circuit. If there is a choice, I suggest a modern flush type, *fig. 75*, which is mounted on a metal wall box you sink into the plaster.

These boxes are about $\frac{5}{8}$ in. deep and most of them now have an earthing terminal, *fig. 76*. Some also have adjustable lugs so that a switch can be aligned if the box is mounted slightly out of square.

The hole for the switch box can be made as described for a socket outlet in chapter 9, but to a shallower depth, of course. The cable to the switch goes through a grommet in a knockout in the metal box. Make sure about 1 in. of the cable's outer sheath enters the wall box.

FITTING THE ROSE

To fit a modern loop-in type of rose, take off the cover and make a hole for the cable by removing a plastic knockout from the base of the rose.

Being fully enclosed, modern roses do not need to be mounted on a pattress. They can be screwed direct to the ceiling joists or to a length of timber firmly fixed between the joists. Older type roses which have no backs must be mounted on a pattress, *fig. 77.*

JOINT BOX METHOD

To add a light to a joint box system is simple. Fix your new joint box in a suitable position. From an adjacent existing box run a suitable length of 1·0mm.² pvc two core and earth pvc sheathed cable to the new box. Connect the live and neutral conductors to their appropriate terminals in each box.

If there is no earth wire used in the existing box, the end of the earthing conductor of your new cable can simply be left in the box, but enclose its exposed portion in the box in insulating tape so that it does not foul the terminals.

In your new box, the other end of the earth conductor should be taken to the earth terminal. Add a length of green insulation sleeving to its bare end for identification purposes.

TWO CIRCUITS

It is desirable that a house should have at least two lighting circuits. No one circuit should supply more than eight (but preferably six) lighting points.

Although the maximum load of a 5 amp lighting circuit is 1,200 watts (equal to twelve 100 watt lamps), some allowance has to be made for fluorescent lamps as for this purpose they are rated at double their quoted rating.

Allowance should also be made for wall lights fitted with two or three lamps, and for any extra lights which may be added to the circuit later on.

When assessing lighting points, the loading of each lamp is taken as 100 watts (even if it is rated below this figure). Each lamp over 100 watts, however, is taken at its actual rating.

For example, if you have only one circuit for lighting, and it supplies four 100 watt, four 60 watt (say for two wall lights), and two 150 watt lamps, the total load is reckoned at 1,100 watts (the 60 watt lamps counting as 100 each).

A system such as this would require eight lighting points, the maximum for a lighting circuit. By having all these points on one circuit, the house would be in darkness if the lighting circuit fuse blew.

USING THE RING CIRCUIT

One answer to this problem, if a second lighting circuit is not possible, is to make use of the ring circuit, not only for operating table and standard lamps, but for the wall lights also.

If, however, two lighting circuits are feasible, it is a good idea to have one for the ground floor and one for the first floor. A third circuit could be used to supply lights for the garage, shed or greenhouse.

DANGER SPOT

A potentially dangerous situation can arise where the landing and hall lights run off the same circuit. When the circuit fuse blows, this area can suddenly be plunged into total darkness. When these lights are controlled by two-way switches (one on the landing, the other in the hall), they run off the same circuit.

One solution to this problem is to fit a separate light on the landing run from a different lighting circuit, or have a table lamp in the hall operating off the ring circuit.

Either method is enough to relieve

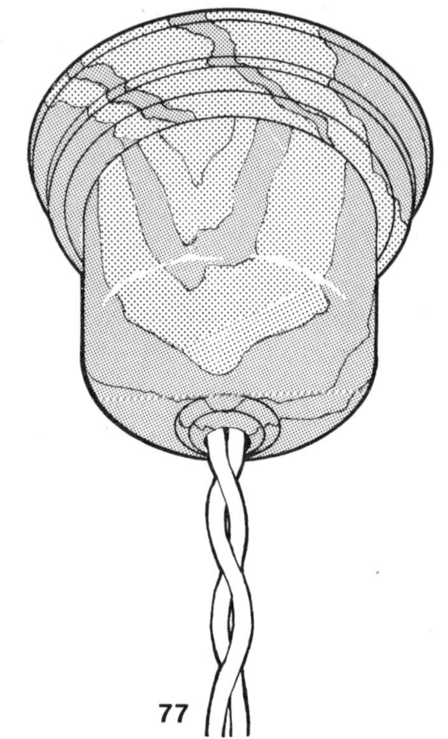

Fig. 77 Ceiling rose and pattress

Fig. 78 Block connectors

total darkness and perhaps prevent a nasty accident.

WALL LIGHTS

The normal method for wiring wall lights is to use the lighting circuit and there are various ways in which this can be done. With a modern circuit properly wired, there should be few problems. Some older houses, however, are wired haphazardly. Additions have been made over the years and the result is often a confused mass of wires.

In my view, a beginner with electrics would be unwise to attempt wiring wall lights from a circuit such as this. No amount of guidance on paper will be of value where an existing circuit defies all the known rules! On-the-spot expert guidance is needed in cases like this.

Personally, if I wanted to fix one or two wall lights in a room, I would do the job just prior to redecorating and I would make use of the ring circuit to do so. For a greater number of lights, however, the lighting circuit should be used.

ALTERNATIVE

Although ring circuits are not intended to supply an entire lighting circuit, they can, within reason, be used to supply the odd wall light or two, or an ordinary light in a remote spot.

One good reason for selecting the ring circuit method for supplying wall lights is that it provides an alternative source of light when a circuit fuse blows on a lighting circuit. Apart from that, the method is fairly straightforward and does not involve a lot of wiring under floorboards or in a perhaps inaccessible loft space.

First, though, a few words in general about wall lights. Many of these are supplied complete with flex, to which is attached a small block connector, *fig. 78*. This enables the light fitting to be connected to the cable conductors.

Most wall light fittings have a built-in switch, but wall lights should be operated also from a master switch. If they are not, the fitting and its flex will always be live.

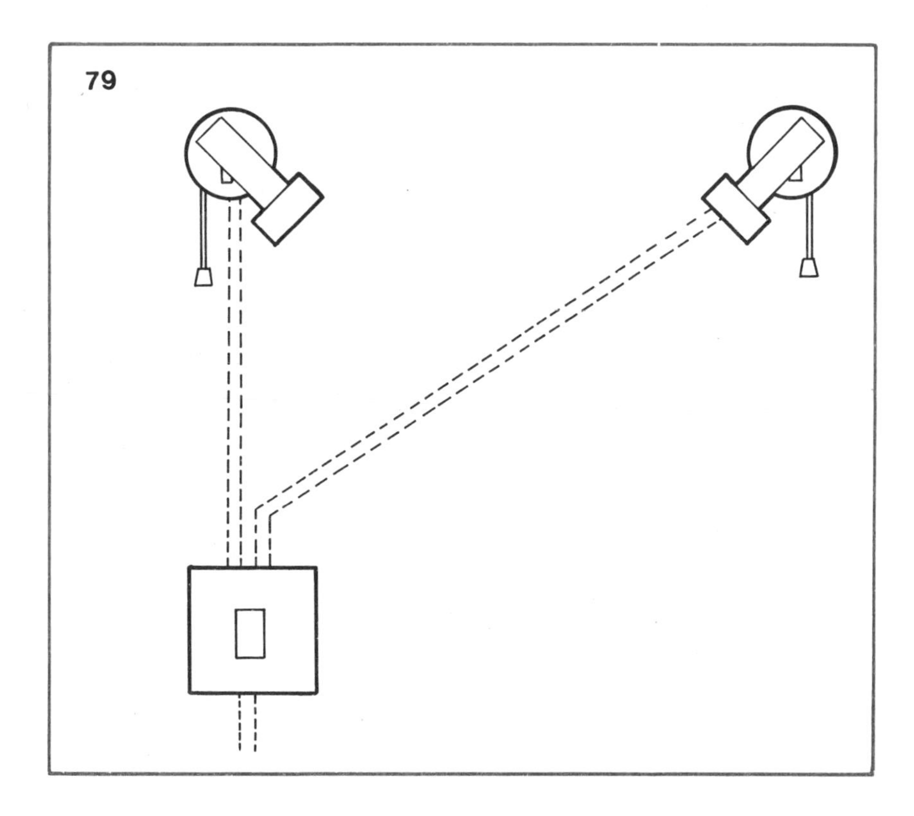

Fig. 79 Wall lights operated from a master switch

FIXING METHODS

There are wide variations in the design of wall lights, particularly in the way in which they are fixed to the wall. In most cases a metal knockout box (sunk into the wall) is needed to accommodate the joint between the circuit cable and the flex of the fitting.

Other types require a conduit knockout box, *fig. 80*; spotlights are one example. But for many types of fitting a narrow architrave metal box, *fig. 81*, is now used.

There are some fittings which do not need a box, such as the Slidona types made by Maclamp. These are fitted with a backplate which is in two parts, *fig. 80*.

The part which contains the cable terminals is fixed to the wall. The other part (attached to the fitting) simply slides into the front part after the connections have been made.

When buying wall lights, study the method of fixing and choose the best one for the situation you have in mind.

TWO METHODS

There are two ways in which the ring circuit can be used to supply wall lights. The simplest is the 13 amp plug and sheathed flexible cord method,

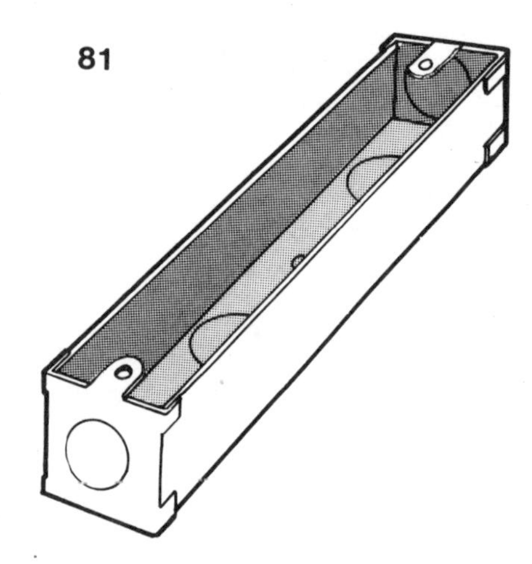

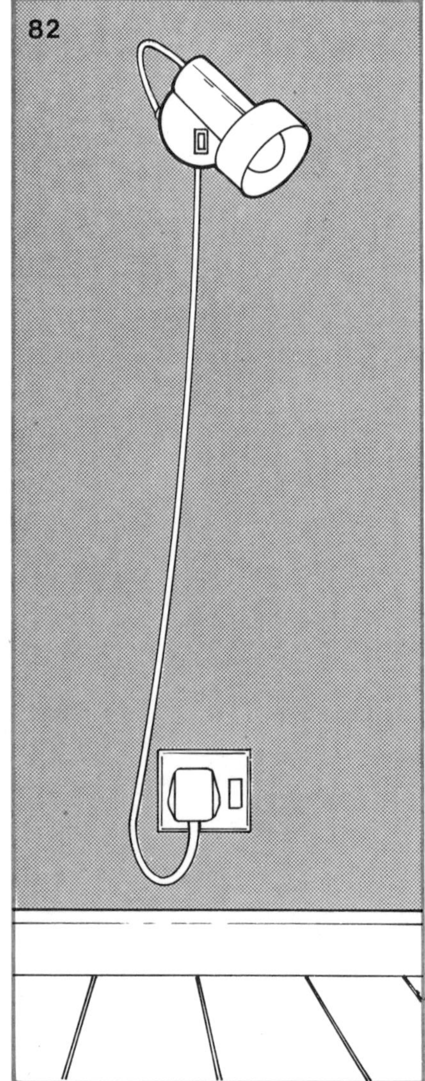

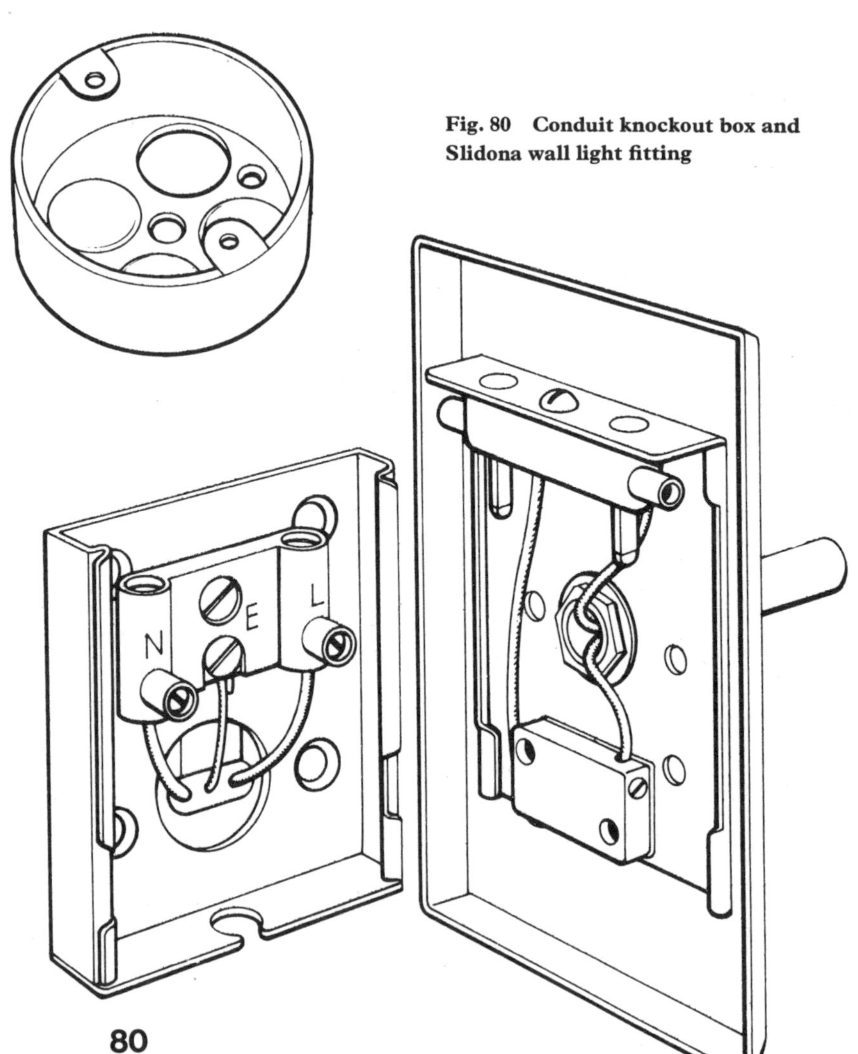

Fig. 80 Conduit knockout box and Slidona wall light fitting

Fig. 81 Architrave wall box

Fig. 82 Wall light supplied by ring circuit plug

Fig. 83 Fixed wiring for wall lights

fig. 82. This can be used for such fittings as spotlights, bedhead lights and pin-up lights such as those made by Conelight.

The sheathed flexible cord is fixed to the wall surface. One end of it goes to the 13 amp plug; the other runs direct into the lampholder of the fitting.

The plug should contain a 3 amp fuse and the socket outlet should be a switched type. This can then act as a master switch for the wall light.

The disadvantages of this method, of course, are that only one wall light can be operated from each plug and socket outlet, and surface wiring (the flexible cord) will be visible. It should not be buried under plaster.

FIXED WIRING METHOD

The other method is shown in *fig. 83.* This is called the fixed wiring method. It involves breaking into the ring circuit at a suitable point. One of the easiest places to do this is at the terminals of a socket outlet which is not a spur already supplying socket outlets. See Adding an Outlet on page 43 onwards.

From the socket outlet terminals, run a short length of 2·5 mm.[2] twin and

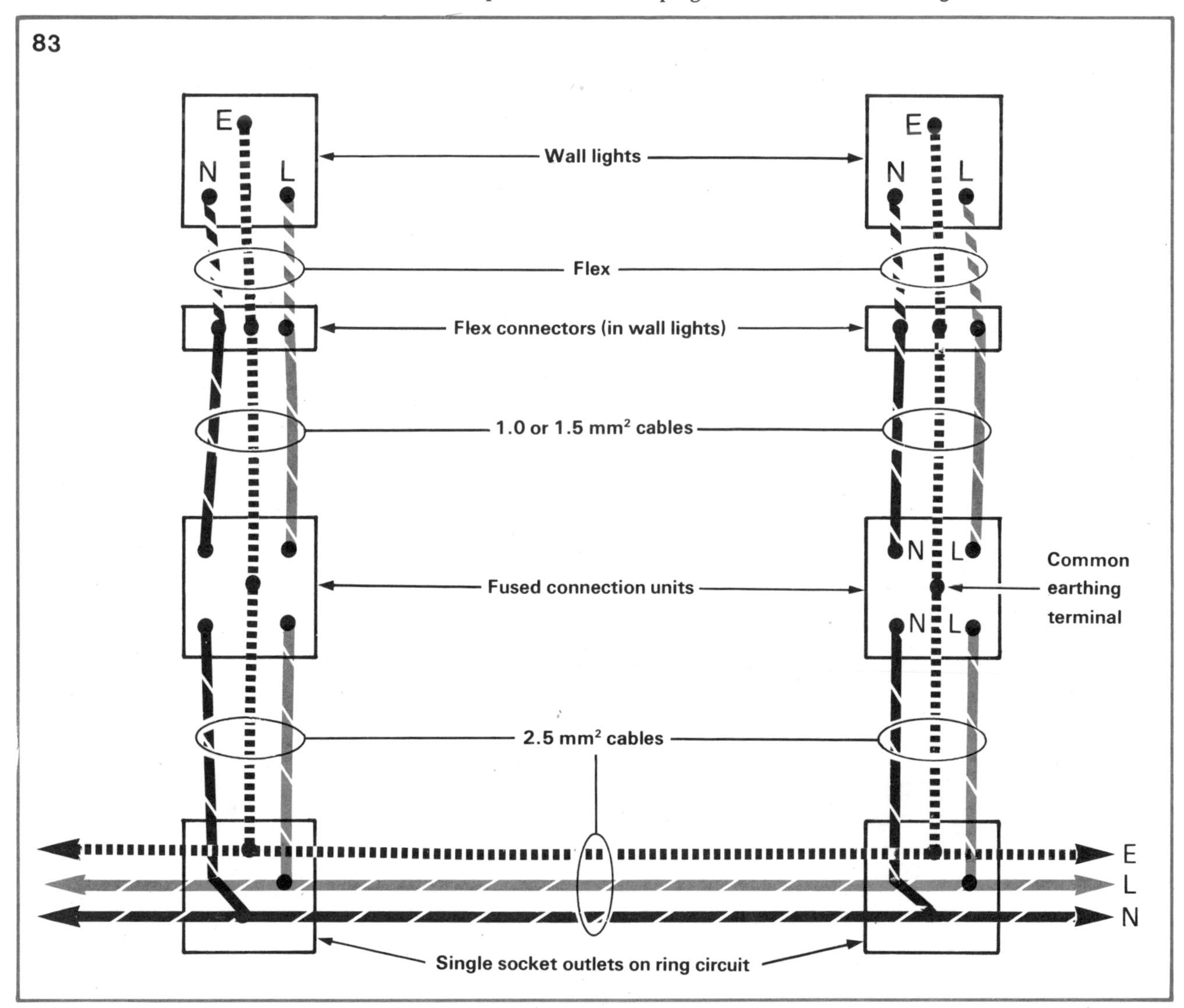

earth pvc sheathed cable to a switched fuse connection unit. This used to be called a fused spur unit or fused spur box.

From this unit run a length of similar cable, but size 1·0 mm.² or 1·5 mm.², to the wall light. It does not matter which of these two size cables you choose. The fused connection unit's switch can be used as the master switch for the lights and as their local switch if they have no built-in switches.

Terminal connections for these units may vary slightly according to make. Generally speaking, there is one common earth terminal to which the circuit earth continuity conductor is connected. The earth continuity conductor of the 1·0 mm.² or 1·5 mm.² cable is also connected to this terminal.

One set of live (L) and neutral (N) terminals on the unit is sometimes marked MAINS. To these go the live and neutral conductors of the cable from the socket outlet on the ring circuit.

The other set of L and N terminals is sometimes marked LOAD. To these terminals go the live and neutral conductors of the 1·0 mm.² or 1·5 mm.² cable to the wall light fitting.

Some connection units are fitted with a hole in the front plate. This is the outlet for the flex from a fixed appliance. You don't need this type for wall lights as the flex from the light will be housed in the wall box and connected to the cable there.

The design you need is shown in *fig. 85*.

The fused connection unit is fixed to a metal knockout box in the same way as described for a socket outlet earlier on.

Do not attempt to wire up wall lights *direct* from the terminals of a socket outlet (or from elsewhere on the ring circuit) without using a fused

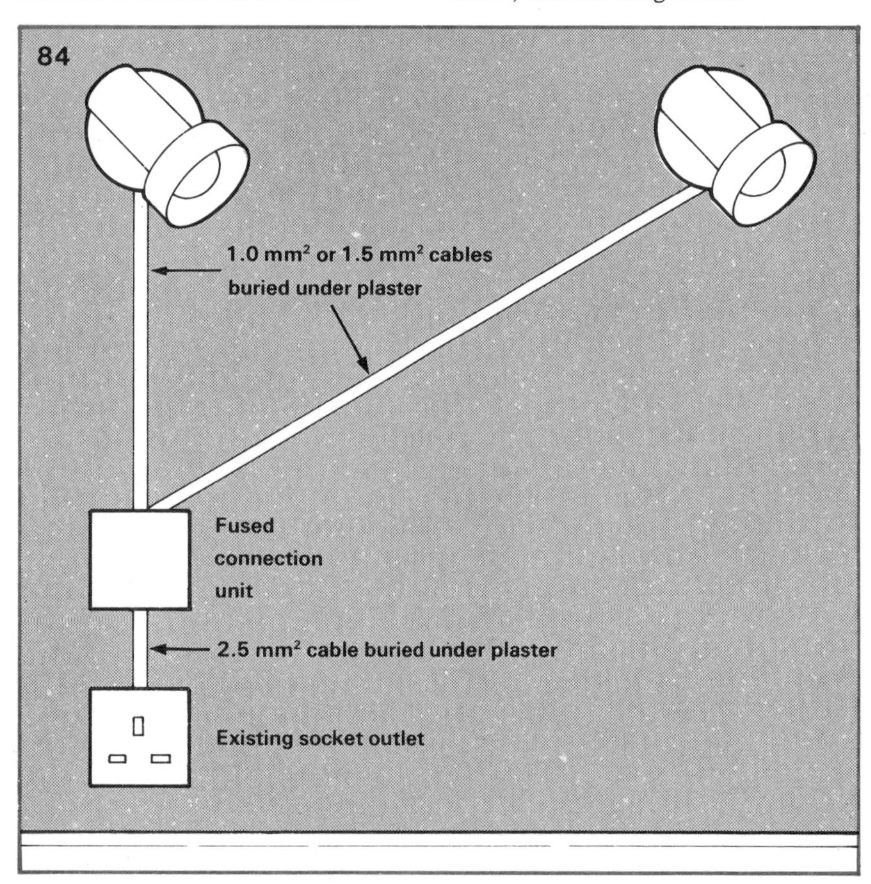

Fig. 84 Two wall lights supplied from ring circuits

Fig. 85 Switched fused connection unit with pilot light

Fig. 86 Fused connection unit wired in a bell circuit

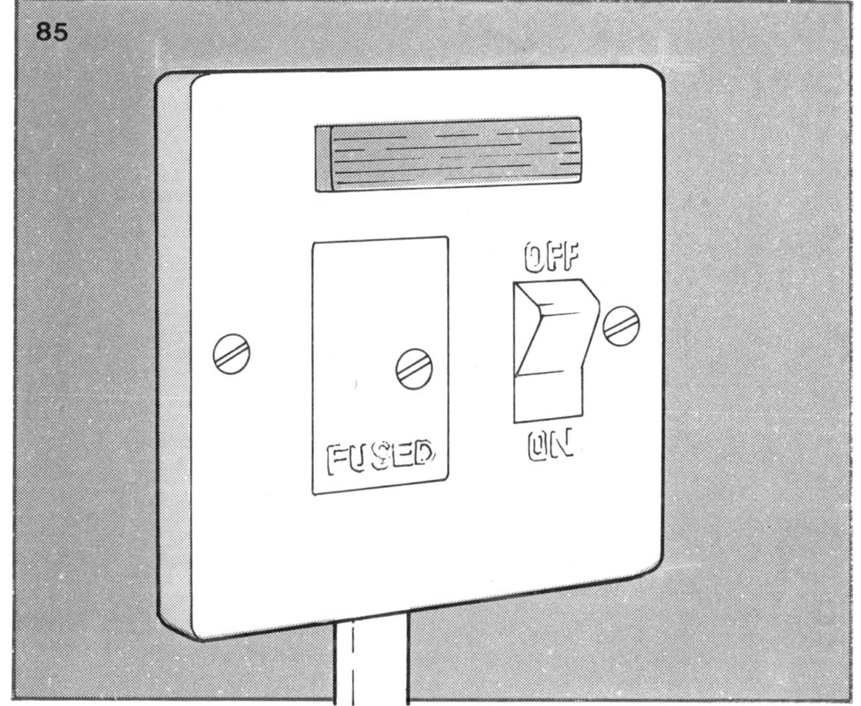

connection unit. This unit should be fitted with a 3 amp fuse which protects the wall light circuit.

More than one wall light can be wired from one unit, but if this is done, the master switch on the unit will control all the lights. So unless the lights have their own switches, all the lights will be off or on at the same time. One way to keep wiring to a minimum is shown in *fig. 84*.

BELLS FROM THE MAINS

While on the subject of fused connection units, let me return for a moment to house bells, already dealt with on page 28.

A fused connection unit can also be used if you want to wire up an electric bell from the mains. The transformer is simply connected to the fused connection unit and the unit to the ring circuit as described above and illustrated in *fig. 86*.

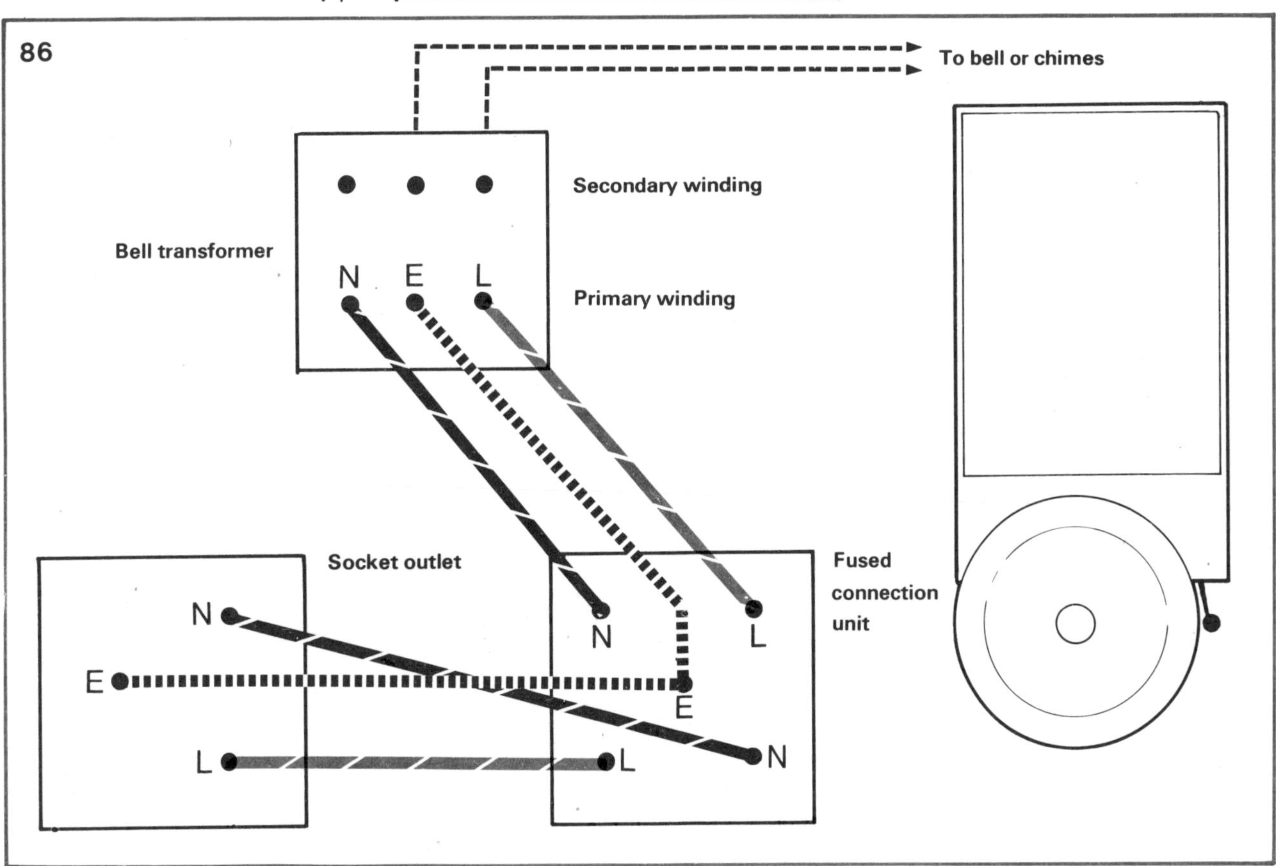

Switches

Old-type switches fitted in rooms with a modern decor are an eyesore. Today's switches are neat, attractive and unobtrusive.

If your wiring installation is old but sound, replacing ancient switches with up-to-date flush types is a job well worth doing.

Probably the ugliest arrangement is an old tumbler switch, *fig. 87*, mounted on a block of wood screwed to the wall or perhaps fixed to a wooden box which is sunk flush into the wall plaster.

OBSOLETE

These methods of mounting switches are obsolete. If you want to replace a tumbler switch mounted as described, here is how to do it – after turning off the current, of course.

If the switch wire running from the ceiling is buried in the wall, the new switch can be either a flush or a surface mounted type, *figs. 88* and *89*. But if the switch wires run in a length of conduit fixed on the wall surface, the new switch will have to be a surface mounted type.

The alternative is to hack away the plaster and possibly some of the brickwork so that both the switch wires and the conduit can be buried under the wall surface.

If the conduit is to stay on the wall surface, the switch will have to be mounted on a suitable size plastic pattress box, *fig. 90*. A hole will need to be made in the top of the box to accept the conduit.

After removing the old switch and its wooden mounting block or box, examine the ends of the cable. If the installation is very old, the ends of the conductors may be coated in verdigris. Clean these up or, if there is enough wire to spare, nip off the ends and prepare clean ends for the terminals of the switch.

If the cable is housed in conduit under the wall plaster, fit a plastic or rubber bush on the end of the conduit to protect the sheath of the cable from chafing.

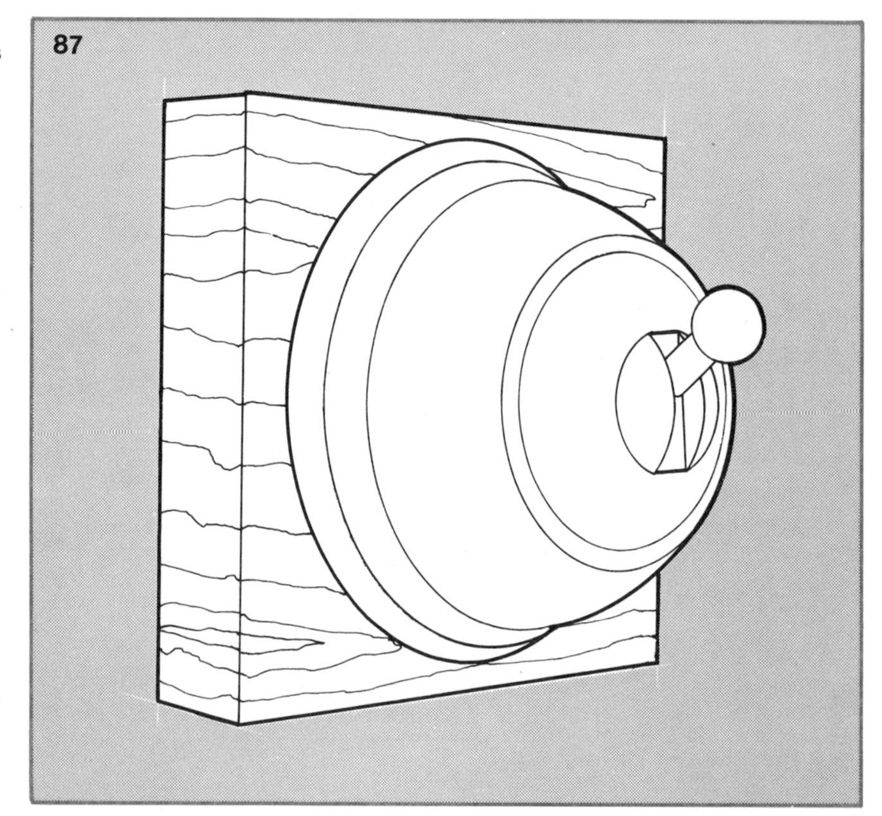

87

88

MOUNTING THE SWITCH
If your new switch is to be flush-mounted, get a plaster-depth steel knockout box which is fitted with an earthing terminal. Cut a hole for the box in the wall and fix as described for a socket outlet on page 51 but to the correct depth.

For surface mounting, the plastic pattress box should be fitted with an earthing terminal. Part of the base of this box is thin plastic, which is knocked out to make a hole for the cable.

Place the box over the old switch position and fix it to the wall with screws into plugs. Make sure it is level. Take care to get the switch the right way up. Most of them are marked TOP on the casing which contains the switch mechanics.

Connect the conductors to their terminals. In an old installation there will be no earth wire at the old switch to connect to the earthing terminal on your new switch box.

Strictly speaking, an earth continuity conductor should be run back from the switch box to the consumer unit or other convenient earthing terminal on the circuit. If this is not possible, the earthing terminal on the new switch pattress can be left empty until the lighting circuit is completely rewired.

TWO-WAY SWITCHES
If you want to incorporate two-way switching in your lighting circuit, for example, in a hall, you will need two-way switches which are fitted with three terminals plus, of course, wall boxes or pattresses fitted with an earthing terminal.

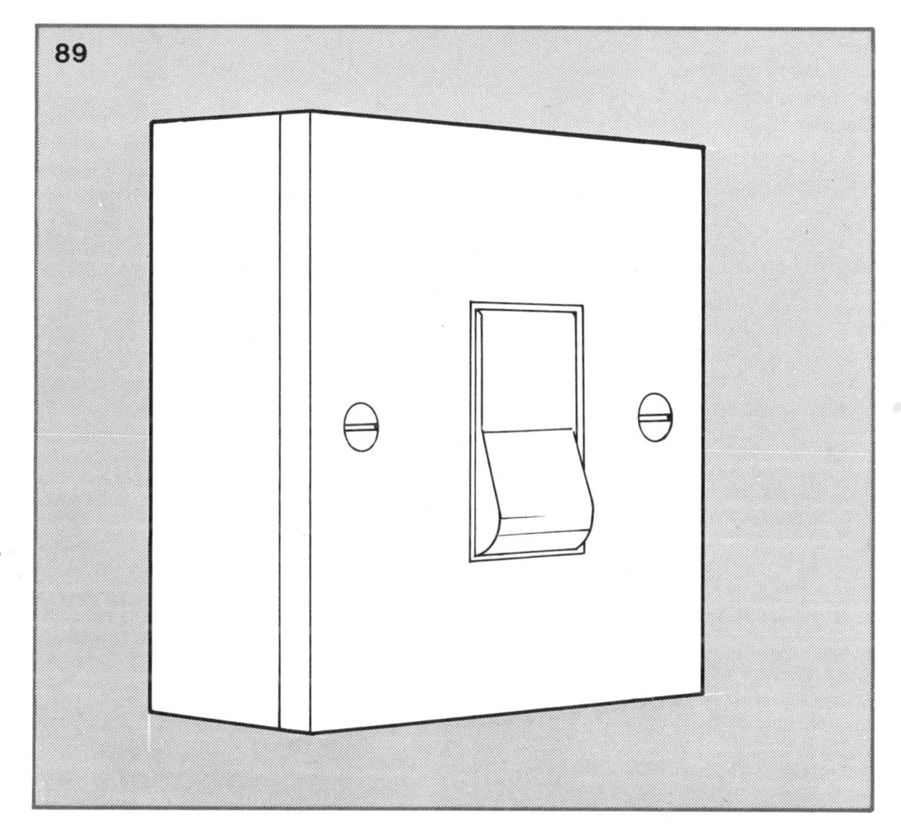

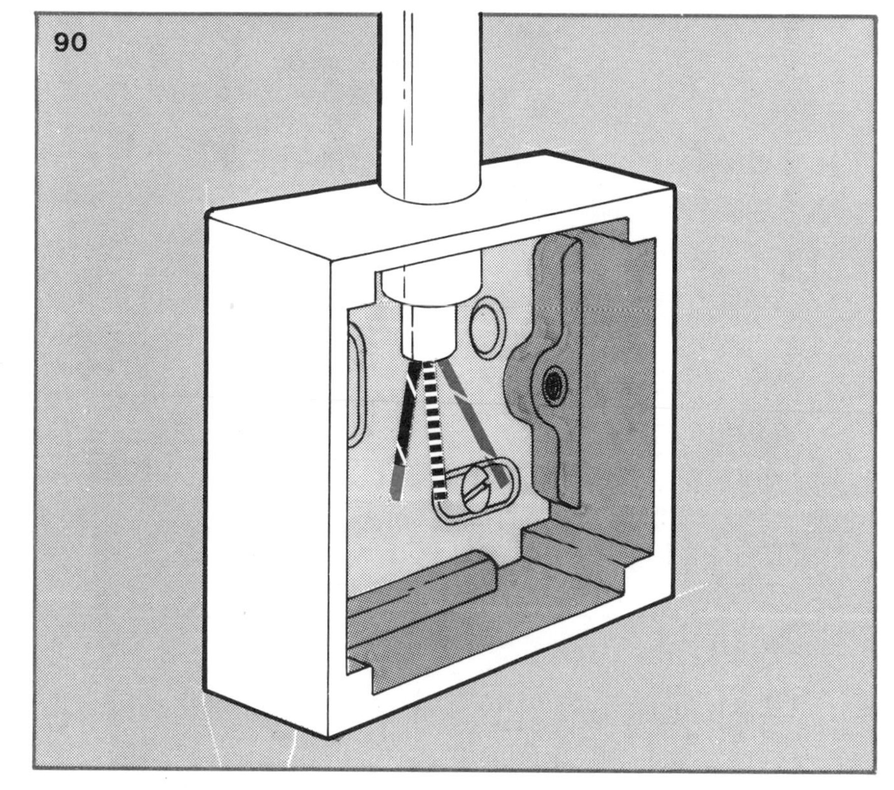

Fig. 87 Tumbler type switch

Fig. 88 Modern flush switch

Fig. 89 Surface mounted switch

Fig. 90 Pattress box for switch

In a two-way switching system, a light can be switched off or on from two positions. For example, one switch could be fitted just inside the front door and the other at the opposite end of the hall.

STRAPPING WIRES

Both switches are connected by what are known as strapping wires. All the cables used in such an installation should be 1·5 mm.², but if your old switch is wired in the imperial size seven strand cable, this can remain. The metric size cable can be used with it in any circuit, provided its size is similar.

The new two-way switches will have one terminal marked Common. The other two terminals are usually marked L1 and L2.

The switch wire to the old switch should be taken to the L1 terminal of the new switch. The wire running from the live terminal of the ceiling rose to the old switch should be connected to the L2 terminal on the new switch, *fig. 91*.

If you have an earth wire at the old switch, which is unlikely, this should now be taken to your earthing terminal on the switch wall box or pattress.

All you have to do now is to link the first new switch with the second using three core and earth cable. This is simple, as *fig. 91* shows. L1 is strapped to L1; L2 to L2; and earth to earth. The other wire of the cable joins the two Common terminals of the switches.

Fig. 91 Wiring two-way switches

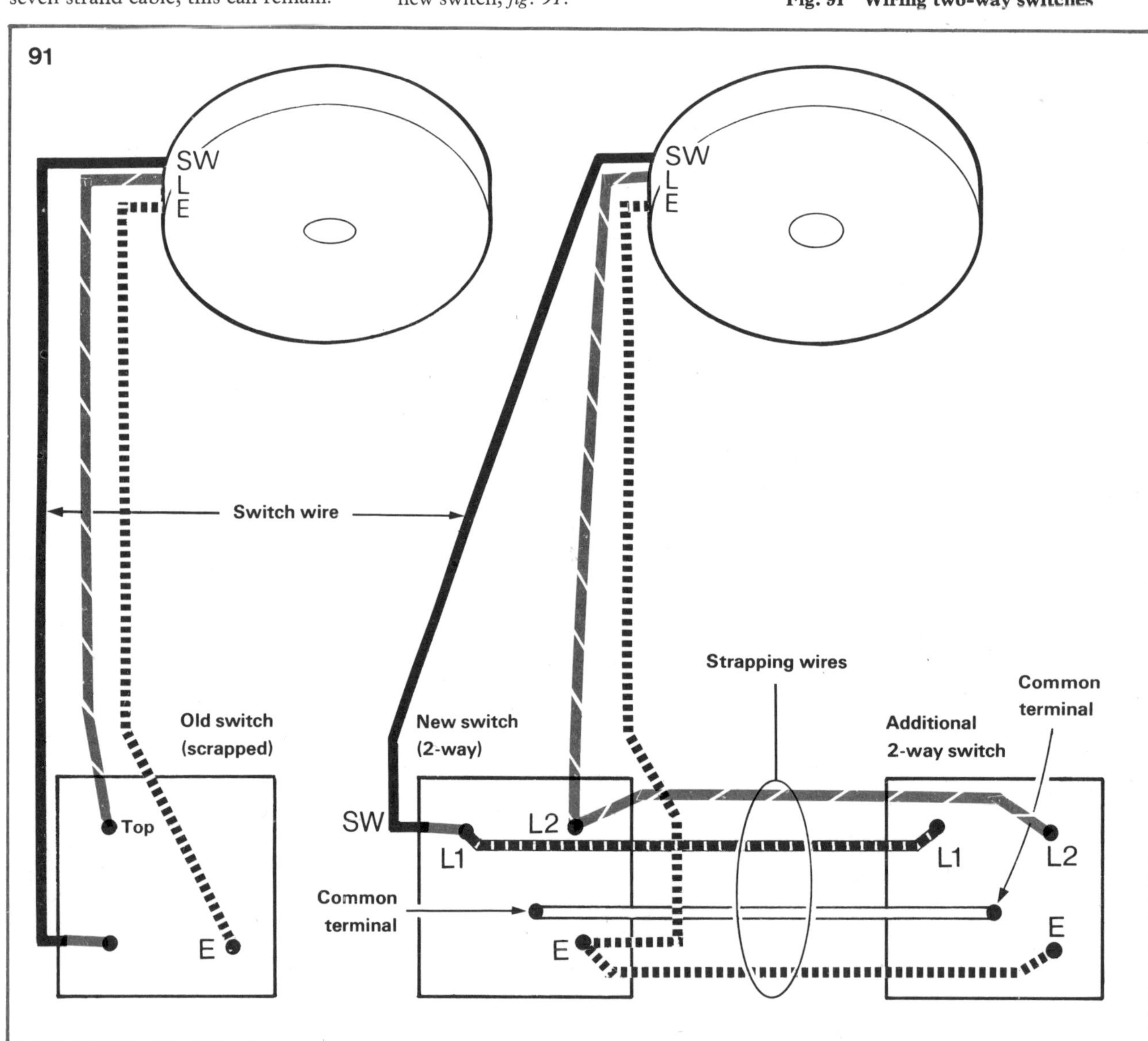

CONTROLS

When buying switches, you may be asked whether you want dolly or rocker operated types. This refers merely to the type of finger control on the front plate of the switch.

A rocker switch, *fig. 92*, operates with a rocking action as it responds to the touch. The more common dolly control, *fig. 93*, simply moves up and down in the more conventional way. Both types will, of course, fit any wall box designed to accommodate switches for domestic use.

Take care when fitting one-way switches to get the switch the right way up. Most have the word TOP clearly marked. If the switch is mounted upside down, the dolly will be down when the light is off and up when the light is on. This could be dangerous!

For example, when putting in a new lamp, one's first reaction is to push the dolly up to switch off the current. If the switch were reversed, the action of pushing the dolly up would be to turn the current on.

A SHOCK

Therefore, the lampholder would be live and potentially dangerous. Even if your fingers did not come in contact with the lampholder contacts, the fact that the bulb would light up immediately it was inserted would in itself be a minor shock to some people.

With two-way switches fitted with dolly controls, each dolly's position will depend on the position of the other: when one is up, the other is down. Therefore, to be on the safe side when inserting a new lamp, turn the current off at the main switch.

Fig. 92 Dolly operated switch

Fig. 93 Rocker operated switch

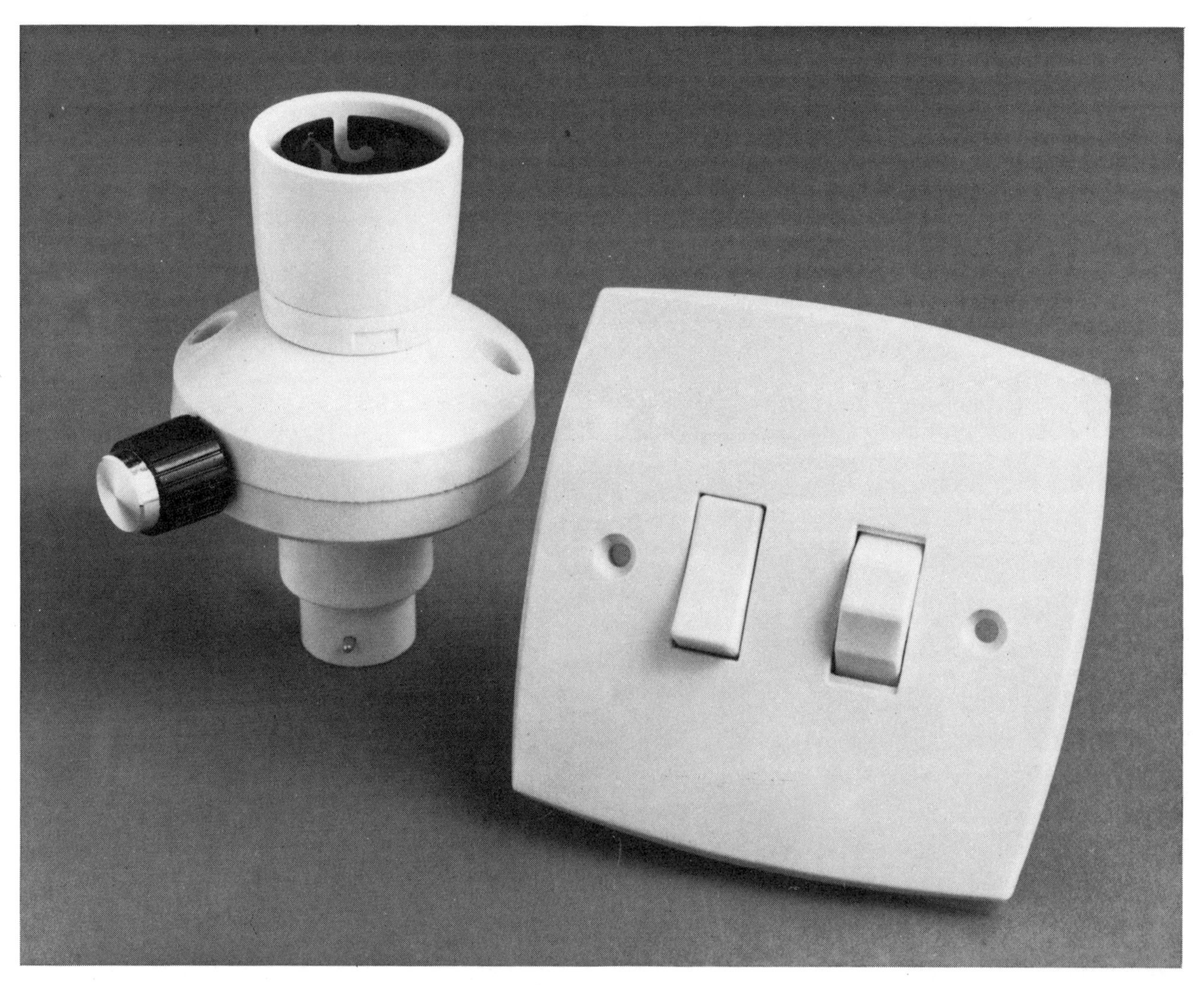

DIMMER SWITCHES

Uses for a dimmer switch, *fig. 94*, in the home are many and varied. Three popular situations are in a child's bedroom (where it can be a substitute for the old-fashioned nightlight); for viewing television in a subdued light; or in the sick room.

There are several types on the market and it is very easy to make an unwise choice. Shop around until you find a type which can be substituted for an ordinary switch *without involving any extra wiring*. Make sure, too, that you choose one in which the printed circuit at the back is completely enclosed so that no live parts are exposed.

If your existing switch is fitted in a plaster-depth box, check that the dimmer is shallow enough to fit it. Some types need a deeper wall box such as that which houses a socket outlet.

DANGER POINT

Some dimmer switches can reduce the amount of light to nothing and yet still remain switched on. This can be dangerous, because if you decide to change the bulb with the dimmer still switched on, you could get a shock. The circuit would still be live.

There is one dimmer (the Myked) which avoids this danger. The light cannot be dimmed beyond a tiny red glow – a valuable reminder that the light is still switched on.

Most dimmers are easy to fit, but follow the maker's instructions carefully.

Dimmers usually have only two terminals – one is for the live feed; the other is for the switch return wire. So you can fit this type of dimmer in place of an ordinary switch without changing the method of wiring.

The fixing holes of dimmer switches will fit a standard plastic surface or metal flush wall box. If your box is not quite deep enough, you can get a mounting piece which fits between the surface plate of the dimmer and the wall box, *fig. 96*.

Fig. 94 Dimmer switches

Fig. 95 Joint box used to extend cable

Fig. 96 Mounting piece acts as spacer

Turn off the current at the consumer unit before you start to fix the dimmer. Remove the fixing screws from the front of the old switch. Disconnect the two conductors and connect them to the terminals of the dimmer. Screw the dimmer to its box and restore the current.

Don't attempt to fit an ordinary dimmer switch to a fluorescent lighting circuit. Special types of dimmers are available for this purpose.

CEILING SWITCHES

A number of circumstances can call for the installation of a ceiling switch in place of an existing wall switch. Two examples which come to mind are a bathroom fitted with a wall switch, or a room which has been converted to a bathroom and also has a wall switch controlling the ceiling light.

The job is reasonably simple. Turn off the current from the main switch, of course, and remove the old switch. Disconnect the ends of the cables from their terminals.

The cable may run down the wall in conduit. If it does, go up to the attic or the next floor, locate the cable and pull it up.

Make a hole in the ceiling for the new switch position, cut the cable to the length required, and make sure you don't cut it too short!

Pass the cable through the ceiling hole and through the backplate of the ceiling switch. Most modern ceiling switches have their own backplate and therefore need no pattress. The connections are similar to those made at a ceiling rose.

If the switch cable is plastered in the wall instead of being enclosed in conduit, it will have to stay there and be cut from the top. You may then find that the remainder of the cable is not long enough to reach the new ceiling switch.

In this case, simply take the cable to a new joint box fitted at a convenient point, *fig. 95*. Then run a new length of cable from the joint box to the new ceiling switch.

The cable to use is 1·0 mm.² or 1·5 mm.² twin and earth pvc sheathed.

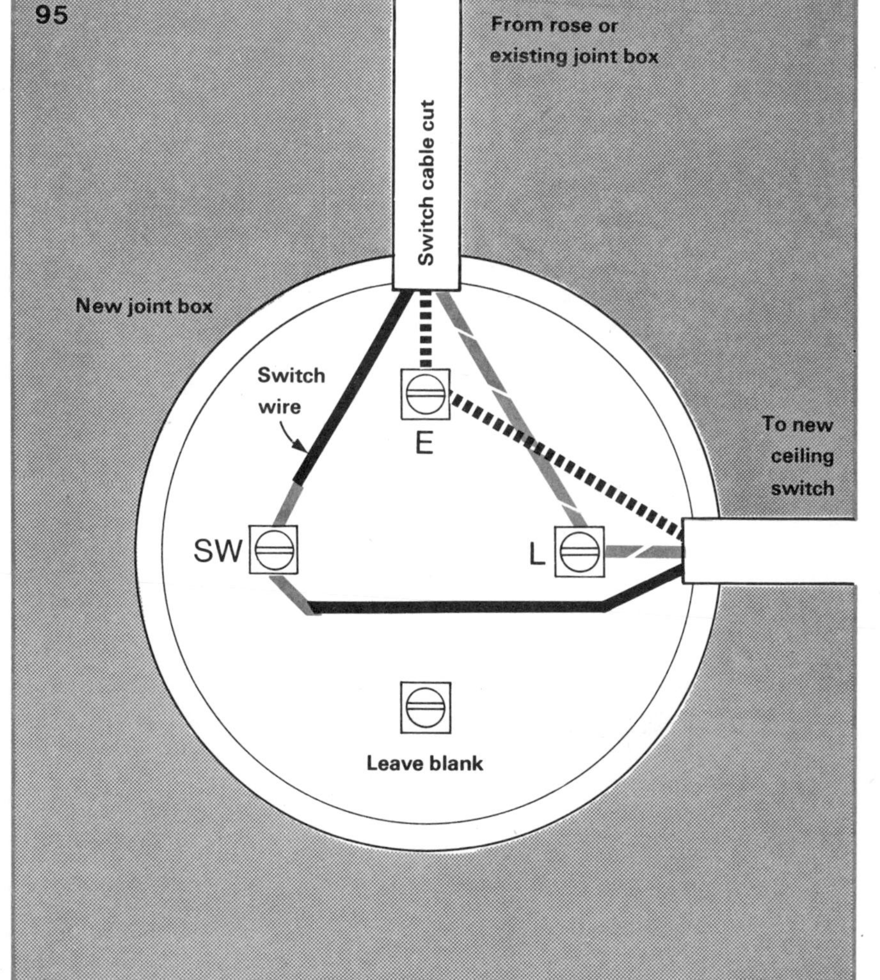

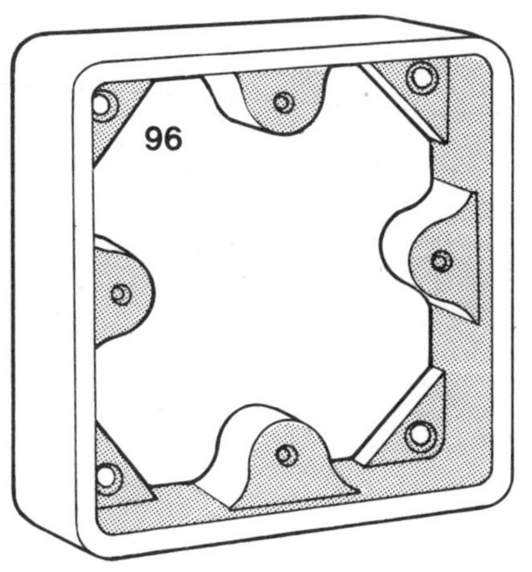

Be Safe!

SAFETY IN THE BATHROOM

Electrical appliances are now used to such a large extent in bathrooms that these rooms can be potential killers. Many accidents happen in bathrooms every year, some of them fatal.

Most of the accidents are caused by using appliances which should never be allowed in the room, such as portable heaters and hair driers.

A primary rule to remember if you want to avoid such accidents is that every piece of electrical apparatus fixed in the bathroom should be properly fitted and correctly used.

Here are a few points worth noting to make your bathroom safe.

1: Fix radiant fires high up on a wall so that they are well out of reach of a person under a shower or in the bath.

2: Make sure that every electrical fitting is firmly fixed and cannot be removed without the use of tools.

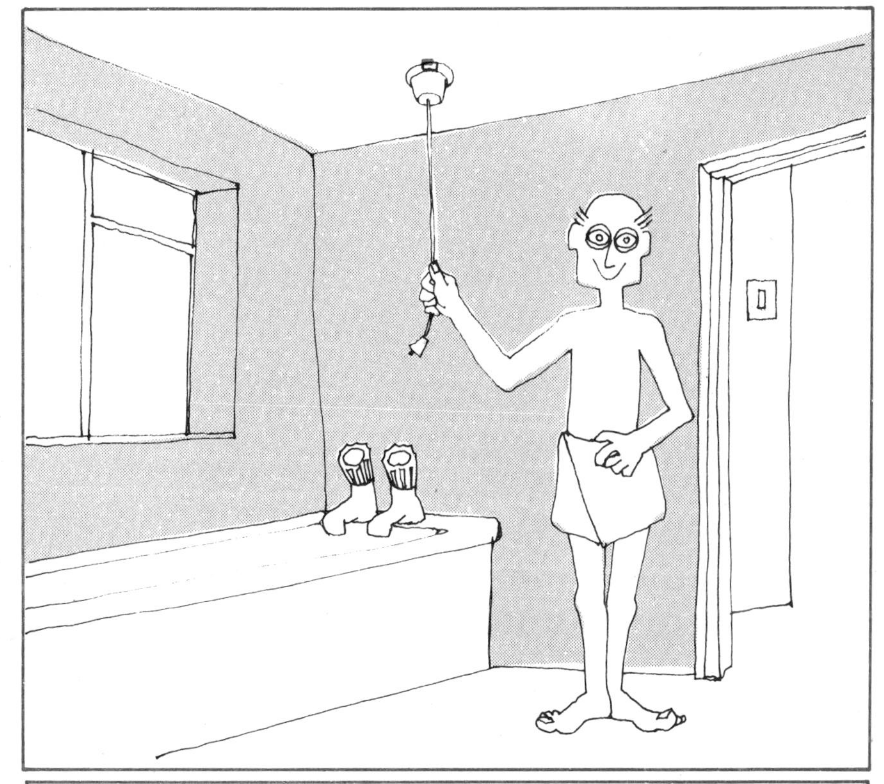

3: See that all heaters and light switches are fitted with pull-cord switches. If this isn't possible the switches should be fitted on the wall *outside* the bathroom door.

4: Cord operated ceiling switches for towel rails or heaters should contain a pilot light so that you can tell at a glance whether they are on or off.

5: If an open reflector type of fire is used, make certain it can be controlled by a separate switch as well as by its own cord-operated switch. The separate (master) switch should, of course, be fitted outside the bathroom.

6: Electric shavers (except battery types) must be used from a properly fitted shaver socket. This must be supplied from a special unit fitted with an isolating transformer. Many strip lights which have a shaver socket are not fitted with a transformer. These must not be used in a bathroom.

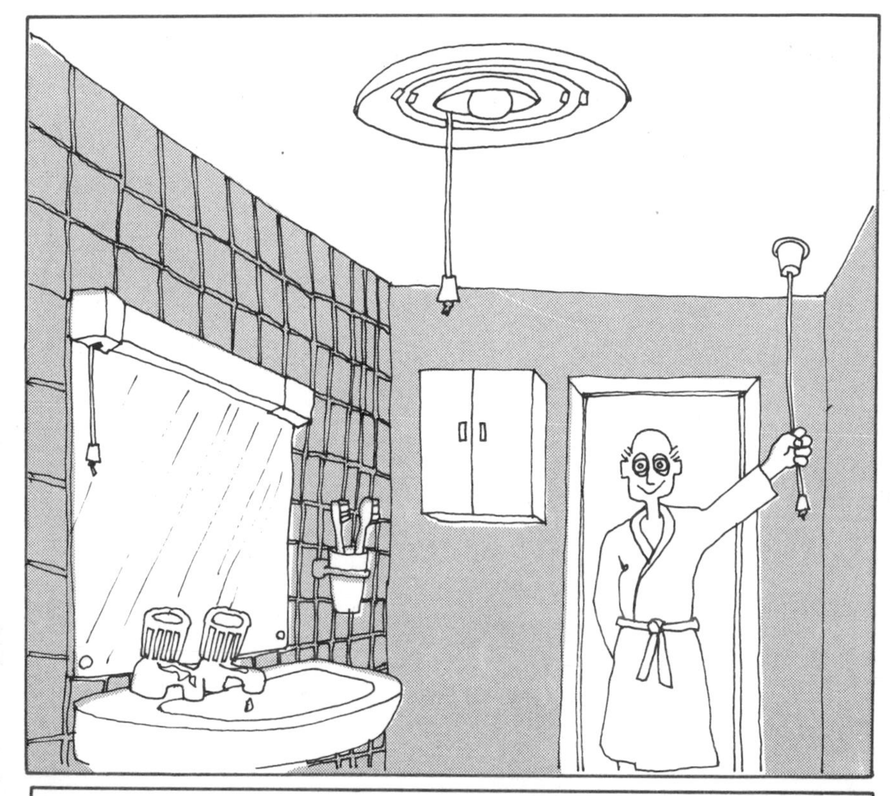

7: Make sure all lights which have a built-in switch are also controlled by a separate switch. Examples are heat/lighting units and strip lights fitted on mirrors.

8: Many people convert a spare bedroom or other room into a bathroom. If you do this, the electrical installation will need to be altered to conform with wiring regulations. This involves removing all socket outlets (or old power points); moving wall switches to outside the bathroom; and using ceiling cord-operated switches. Also, a pendant type of light fitting suspended by flex should be changed and a ceiling type fitting used instead. Steam in a bathroom can damage flex very quickly.

9: Ensure that the lamp bulb cannot be changed by a person who is standing in the bath or under the shower.

10: *Never* use a portable appliance in a bathroom.

11: No switch should be within reach of a person in the bath or under the shower except an insulated cord-operated switch.

12: Never take a television set or radio connected to the mains into the bathroom.

Storage Heaters

The most common form of electric central heating is supplied by storage heaters, *fig. 97*. They operate during off-peak periods, being normally switched on from 11 p.m. to 7 a.m.

During these hours the heaters are charged with heat under a white meter or other off-peak electricity tariff. Electricity is cheaper during these hours, under these special tariffs, than throughout the day when demand for power is high.

Although the heaters give out some heat during their charging period, most of their stored heat is discharged from 7 a.m. to 11 p.m. – the period when you need it most.

Heat output is highest at 7 a.m. when the off-peak period normally ends. During the day, the heat output gradually reduces.

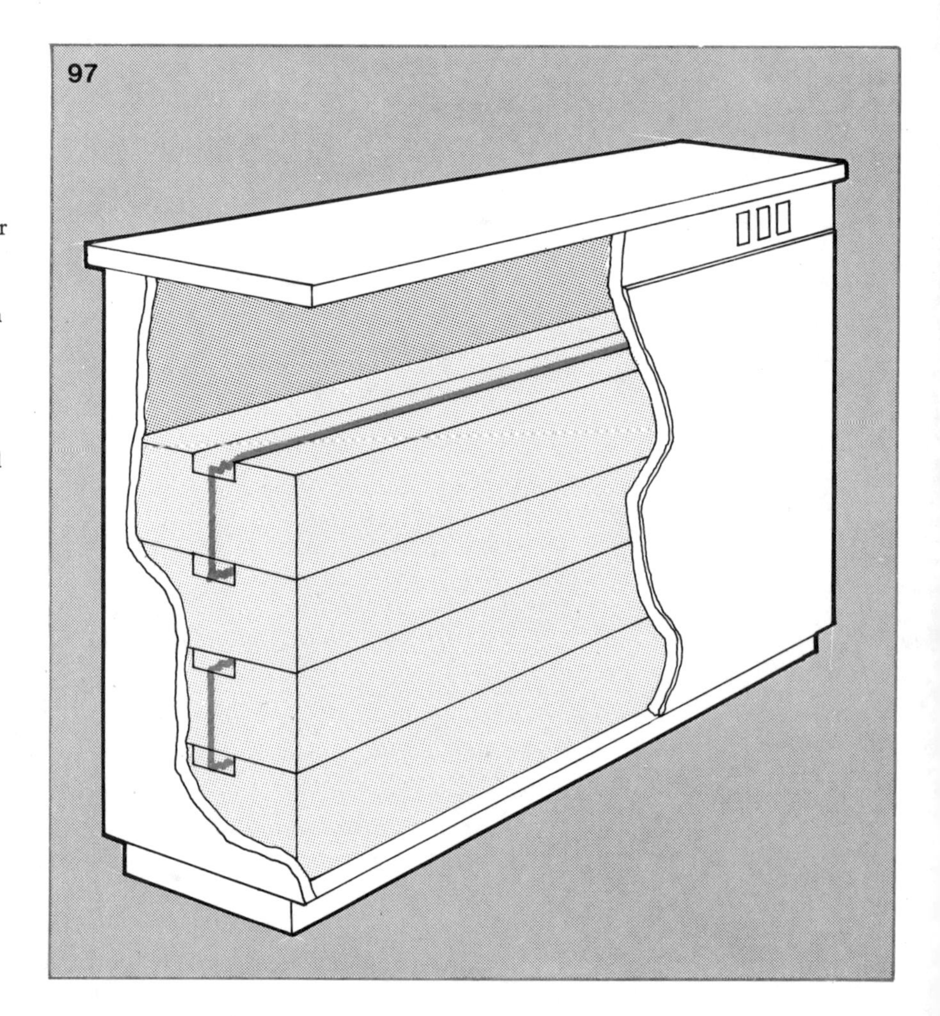

Fig. 97 Storage heater

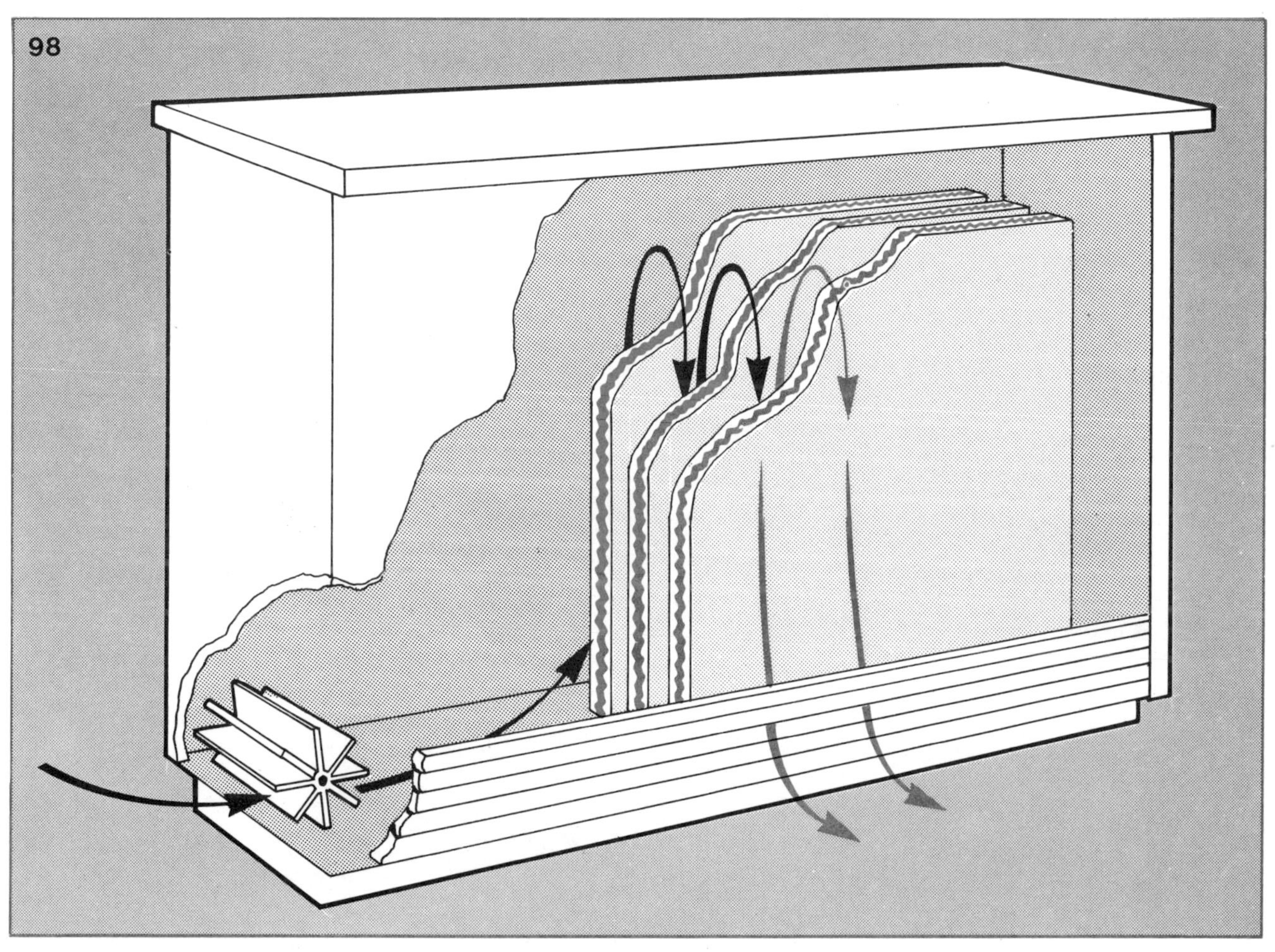

Fig. 98 Fan type storage heater

Storage radiators are the most popular type of heater used for this purpose. An alternative is the storage *fan heater*, *fig. 98*. This is not to be confused with the ordinary type of portable fan heater, of course.

Both types of storage heater come in various sizes (loadings) – storage radiators from $1\frac{1}{2}$ kilowatts to about $3\frac{1}{3}$ kilowatts, and storage fan heaters from 3 kilowatts to round about 8 kilowatts.

OUTPUT

A point to note about these loading figures is that they do not represent the actual amount of heat discharged (output). For example, the largest type of storage radiator ($3\frac{1}{3}$ kilowatts) will produce heat at the rate of roughly 1 kW continuously for most of the day before it gradually begins to reduce its output.

Although it is not possible for you to alter the heat *output*, the amount of heat stored *can* be controlled by an input controller fitted on the heater. This, of course, will to a degree affect the amount of heat which is discharged during the day.

As its name suggests, a storage fan heater is fitted with a fan which drives warm air out into a room through a grille. These heaters are dearer than storage radiators and are normally used only if you want to control the amount of heat generated or to choose the time when the heat is to be discharged. This can be done by using the heater's fan switch. Alternatively, it can be done by having a thermostat or time switch incorporated in the fan circuit.

WIRING STORAGE RADS

Installing storage radiators is a job a competent home electrician can do himself. Wiring up for storage fan heaters, however, is slightly more complicated as two circuits are needed: one is for the heater itself; the other is for the fan.

To take storage radiators first, an important point to remember is that each radiator should preferably have its own separate circuit.

As these heaters are on continuously and run at a reduced tariff rate, they must have a consumer unit to themselves. They are not normally connected to the ordinary consumer unit which supplies the ring circuit and other circuits.

HOW MANY?

If you decide to go in for this type of heating, the first thing to decide is not only how many heaters you need *now*, but whether you are likely to need more later on. Your requirements will dictate the size of the consumer unit.

For example, if you need two storage radiators (or fan heaters) now, the smallest size consumer unit required will be a two-way. But if you also want to run your immersion heater off the same cheaper tariff (discuss this with your electricity board), then you will need a three-way unit.

That will give you two fuseways for the storage heaters and one for the immersion heater. This obviously makes no provision for fitting additional storage heaters later on. Therefore, while you are doing the job, it is well worth while putting in a larger consumer unit now to provide spare fuseways for use later on. You can get units fitted with up to ten fuseways for this purpose.

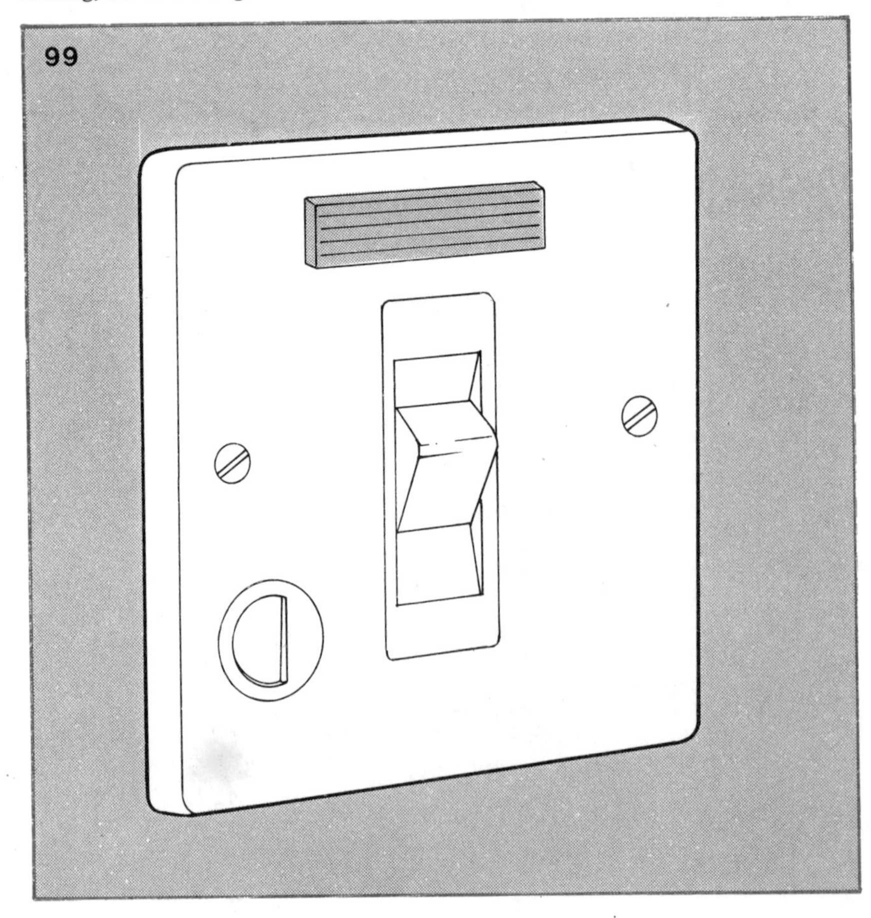

Fig. 99 Double pole switch (20 amp)

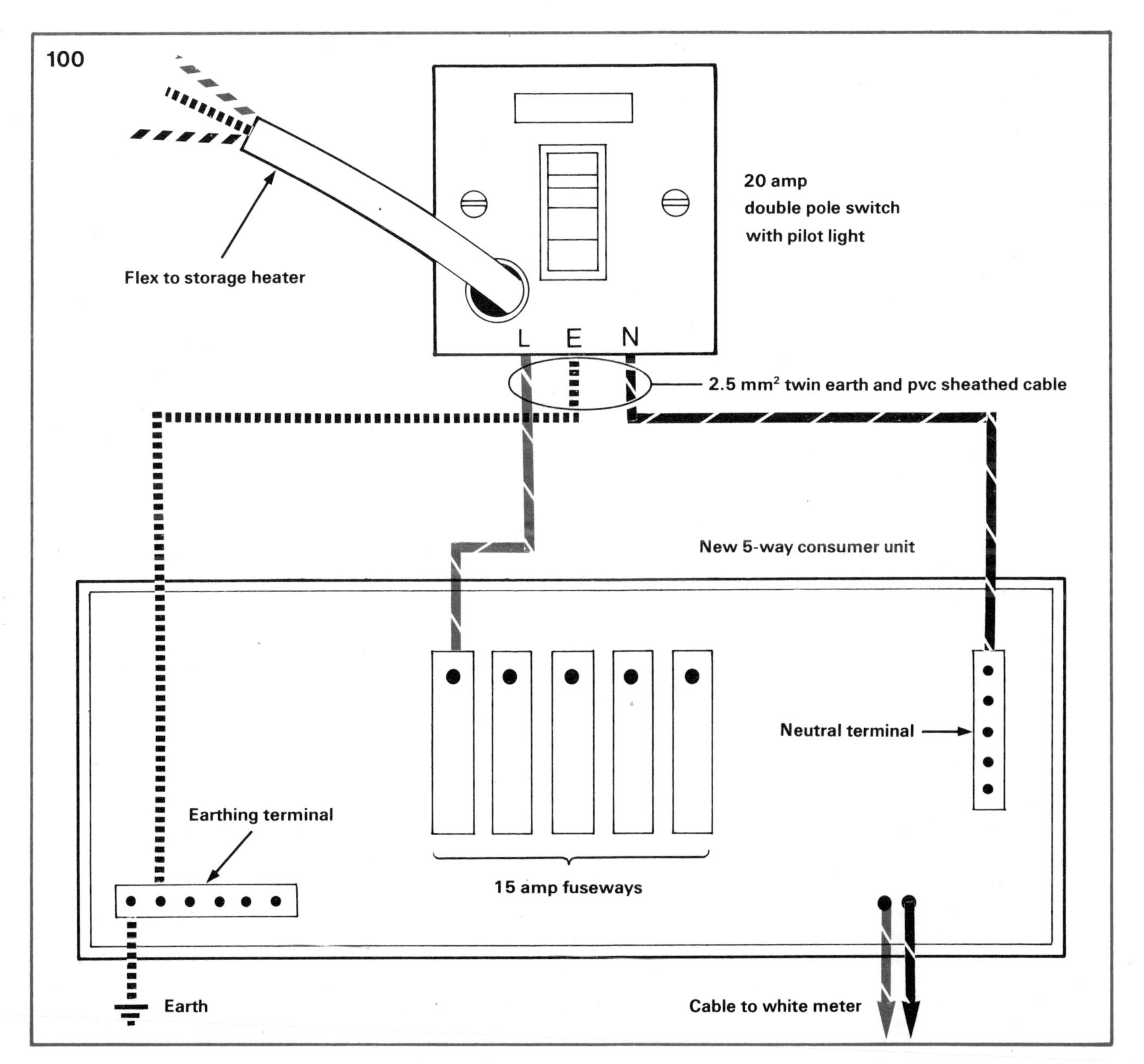

Fig. 100 Five-way consumer unit wired for one storage heater

WIRING UP

To wire up for a storage radiator, you use 2·5 mm.² twin and earth pvc sheathed cable (the same as is used in ring circuits). The rating of the fuse in the consumer unit should be 15 amps.

The best circuit outlet to use for connecting the heaters up is a 20 amp double pole switch. You will need one for each heater. This switch is suitable for all sizes of storage radiators with loadings of 4·8 kilowatts or less.

The double pole switch is fitted into a flush metal or plastic surface pattress type of wall box as used for 13 amp socket outlets. The switch should be the type provided with a cord outlet, *fig. 99*. This provides an entry for the flexible cord which runs from the heater to the switch.

Figure 100 shows a five-way consumer unit wired up for one storage heater. Additional heaters are wired in the same way.

Fig. 101 Wherever warm, dry air exists, a humidifier is desirable to prevent damage to woodwork and furniture

STORAGE FAN HEATERS

The heater section of a storage fan heater is wired similarly to a storage radiator. A 2·5 mm.² twin and earth pvc sheathed cable runs from a 15 amp fuseway in your separate consumer unit, but in this case it terminates at a 25 amp twin double pole switch, *fig. 102*, fixed near the heater. Each heater should have its own 25 amp switch, *fig. 103*.

Fig. 102 25 amp double pole switch
Fig. 103 25 amp double pole switch wired to consumer unit for storage fan heater

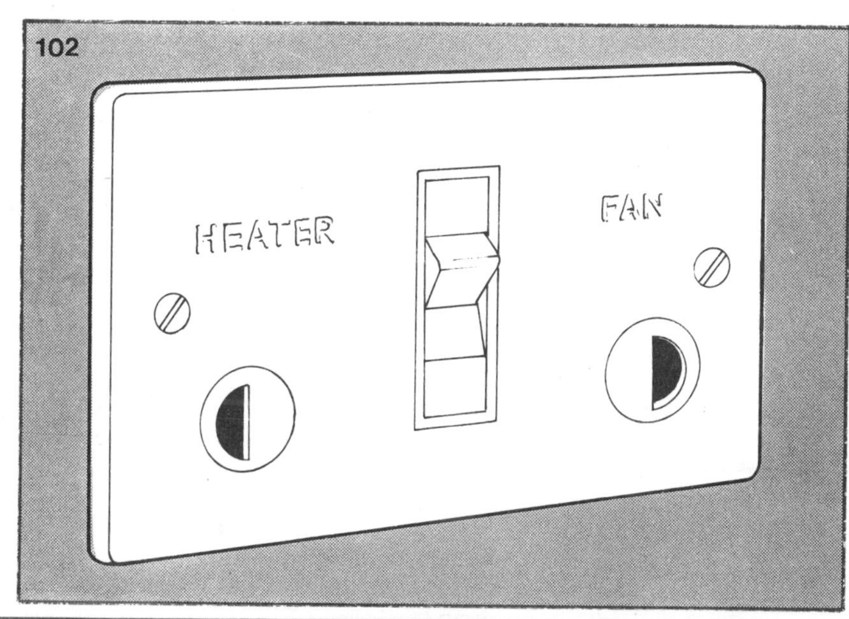

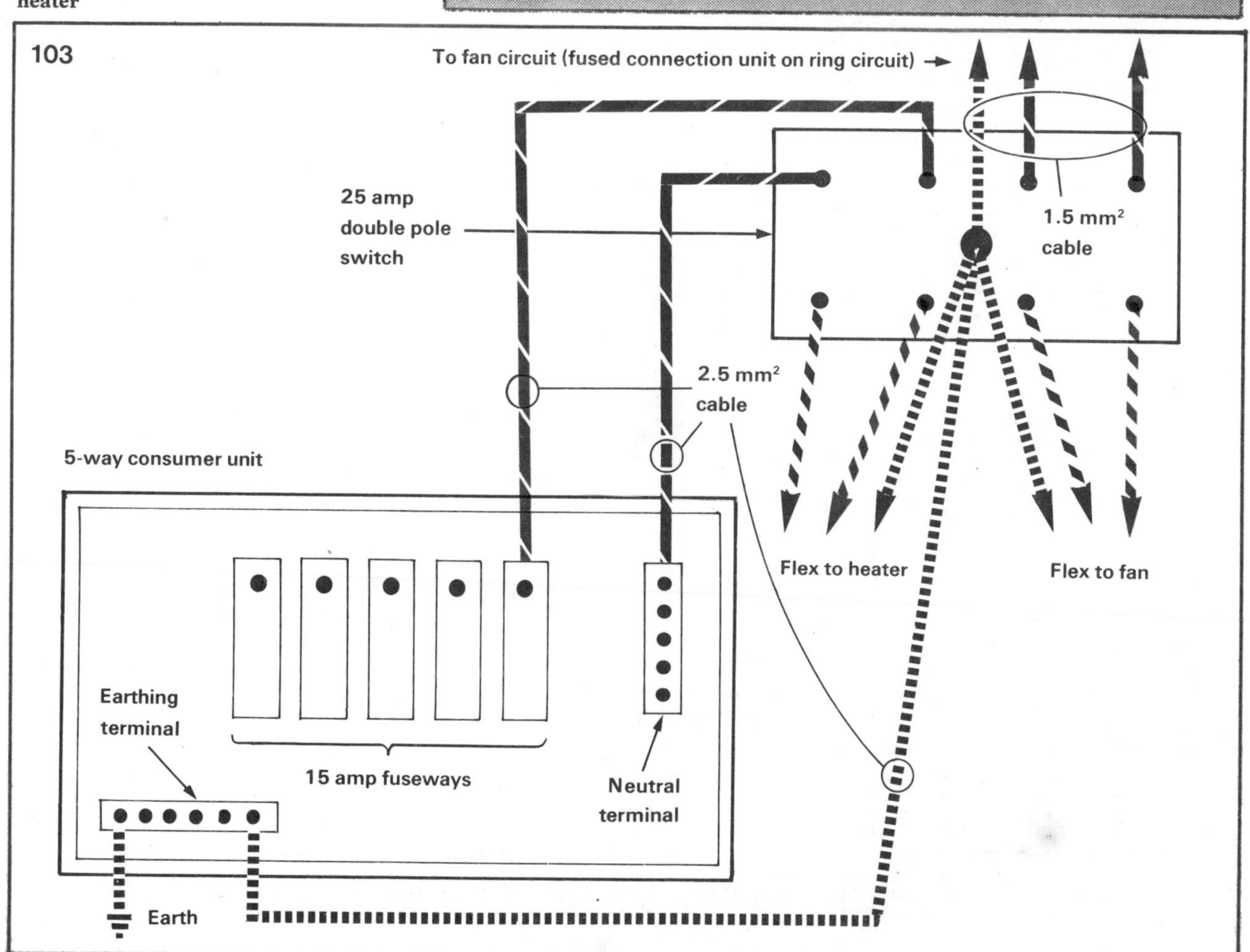

To feed the *fan* with current, you can use your ring circuit. From there you take a 1·0 mm.² twin and earth pvc sheathed cable to the 25 amp twin switch. This cable has to be joined to the ring circuit via a fused connection unit which should be fitted with a 3 amp fuse.

One way to do this job is to fit the fused connection unit next to a convenient 13 amp socket outlet. From the terminals of that outlet, you loop out to the fused connection unit using 2·5 mm.² twin and earth pvc sheathed cable, *fig. 104*.

(Alternatively, you can loop out of the lighting circuit to feed the fan but not, of course, the heater itself.)

The 25 amp twin switch serves a double purpose. One side of it controls the heater; the other operates the fan. It has two cord outlets – one is for the flex to the heater; the other is for the flex to the fan.

This switch is for isolating purposes only and is a linked type operated by one dolly. This switches both the heater and the fan off and on together. The heater itself, of course, has its own built-in switches for fan and heater.

Fig. 104 Heater's fan supplied by ring circuit

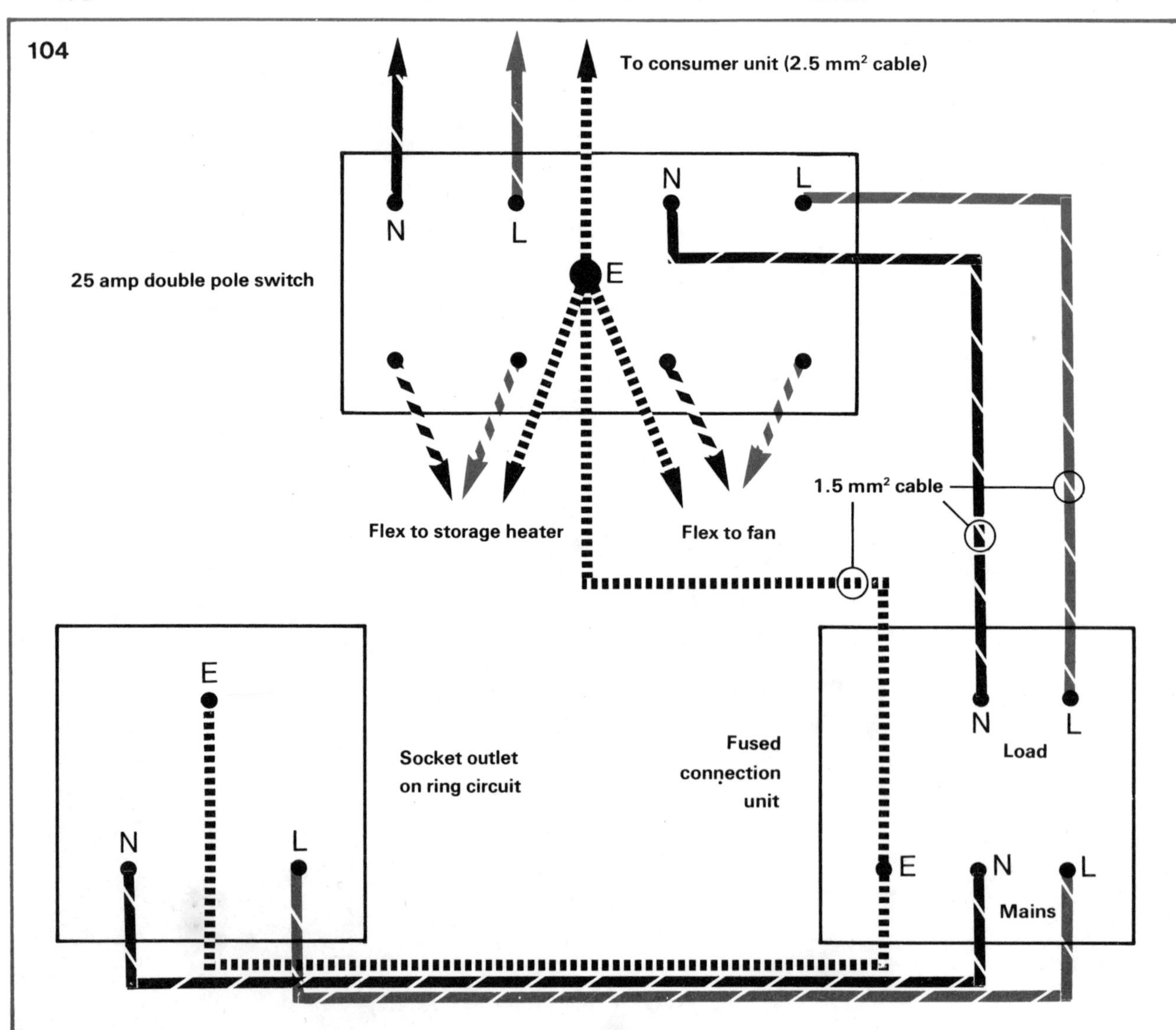

Appliances

Many electrical appliances used in the home can be easily connected to the wiring system by means of a plug fitted to a length of flexible cord. These are called portable appliances and examples include irons, vacuum cleaners, drills, some heaters, hair dryers and so on.

Fixed appliances, however, usually connect to the wiring system by other methods. Some involve installing a new circuit or extending an existing circuit. These appliances include cookers, immersion heaters, towel rails and storage heaters.

To take the cooker as an example. The only appliance which can be connected to a cooker circuit is, in fact, the cooker. On the other hand, its circuit can supply *both* sections of a split level cooker. This type of cooker is operated from a single control panel.

Normal-size cookers are wired in 4·0 mm.^{2} two core and earth pvc sheathed cable. The circuit is 30 amps.

Large cookers which have a loading higher than 12 kW have a 45 amp circuit, however. The size of cable used in this case is 10·0 mm.^{2}. For this type of cooker you need a 45 amp fuse fitted to the consumer unit.

The cable of a normal-size cooker runs from a 30 amp fuseway in the ordinary domestic consumer unit to the cooker's control unit. These units usually consist of a double pole switch for the cooker and a 13 amp three-pin socket outlet for an electric kettle, *fig. 105*.

A double pole switch breaks the current not only in the live conductor, but in the neutral conductor as well.

The units can be flush-mounted or surface-mounted types and you can get them with or without pilot lights.

The control unit for the cooker can be fitted up to 6 ft. away from the cooker itself. The cable which connects the cooker to the control unit is the same size and type as that used from the consumer unit to the control unit, 4·0 mm.^{2} or 10·0 mm.^{2}.

This cable can be buried under the wall plaster and terminated at a cable connector box positioned low on the kitchen wall.

Fig. 105 Double pole switch for cooker

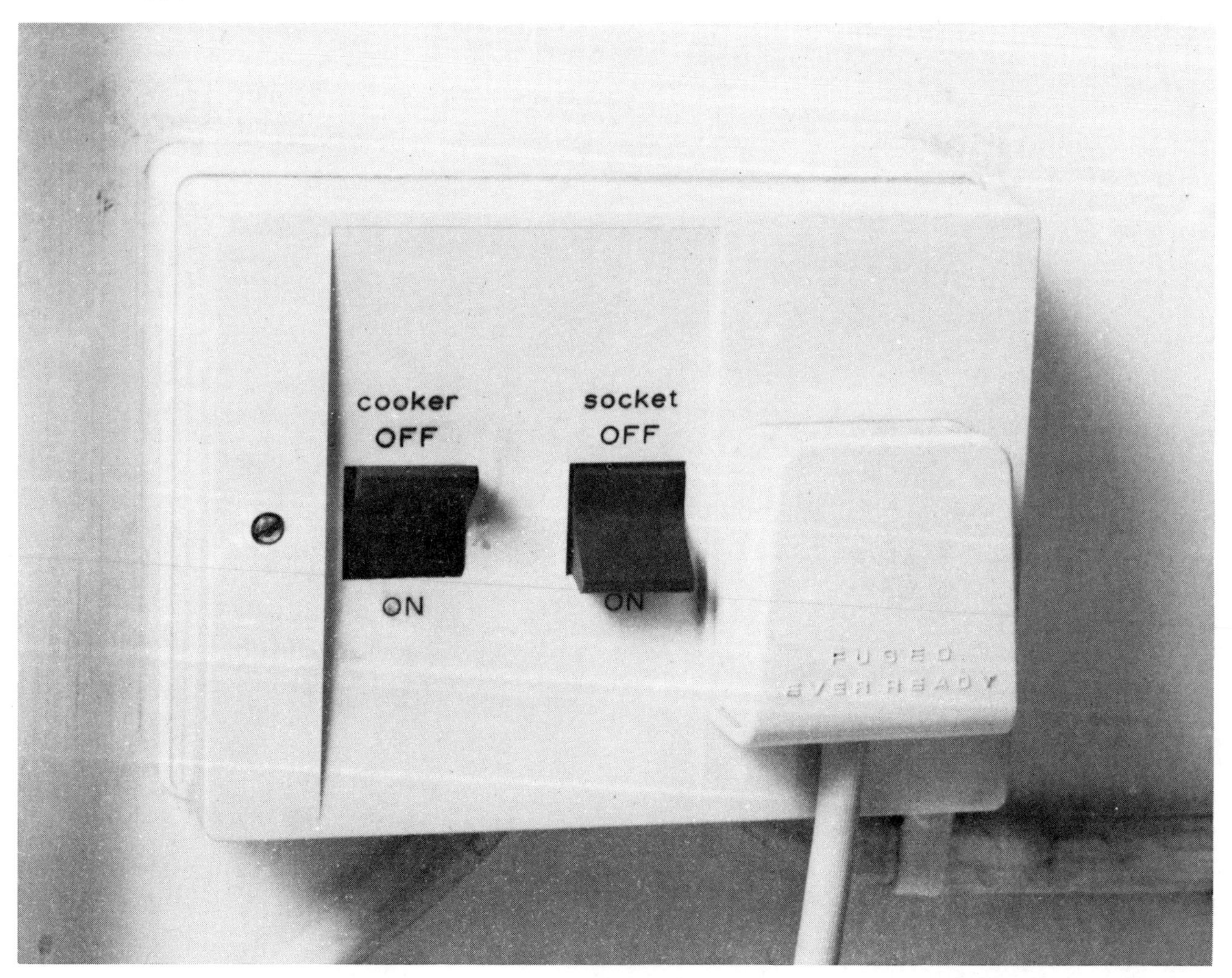

TWO TYPES OF BOX

There are two versions of cable connector box. One is a terminal box, *fig. 106*, in which the fixed portion of the cable and the lead on the cooker are connected up. This allows the cooker to be moved without interfering with the wiring.

The other type is called a through box, *fig. 107*. This type has a cable grip which avoids cutting the cable to make a joint between the fixed portion of cable and the lead on the cooker.

It is possible to supply both units of a split level cooker – the oven and the hob – from one circuit, and to control both with one switch. However, neither of the units should be positioned more than 6 ft. away from the control unit.

You can run separate cables to each from the control unit provided that they are the same size as the circuit cable.

Small table cookers which have a loading of less than 3 kW can be operated from a fused 13 amp plug and socket outlet on a ring circuit.

Fig. 106 Cooker terminal box

Fig. 107 Cooker through box with cable grip

IMMERSION HEATERS

Generally speaking, immersion heaters should not be supplied from a ring circuit but from a separate 15 amp circuit. The reason for this is that an immersion heater is considered to be an appliance with a continuous load of 3 kW. This is regardless of whether you keep it switched on all the time or not.

As the capacity of a ring circuit is only 7·2 kilowatts, to add an immersion heater to it would reduce that capacity by 3 kW.

The heaters are normally supplied by a 2·5 $mm.^2$ two core and earth pvc sheathed cable. This runs from a 15 amp fuseway in the consumer unit to a 20 amp double pole switch with a pilot light, *fig. 108*. The switch should be positioned near the heater and have a flex outlet.

The wiring from the switch to the heater should be three-core 20 amp butyl rubber heat-resisting flex.

Fig. 108 Double pole switch with pilot light

109

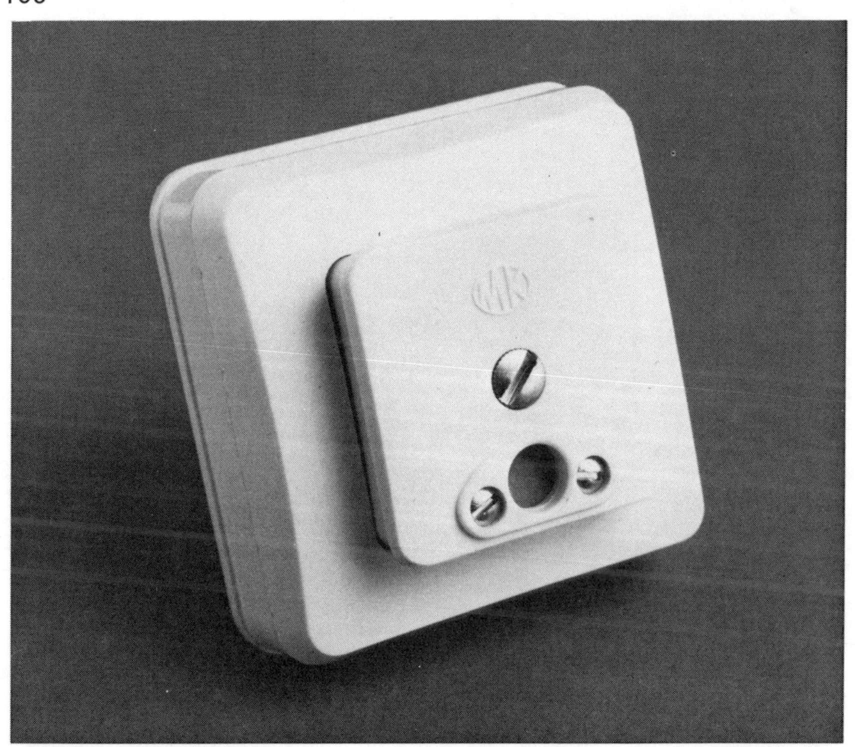

Fig. 109 Clock connector

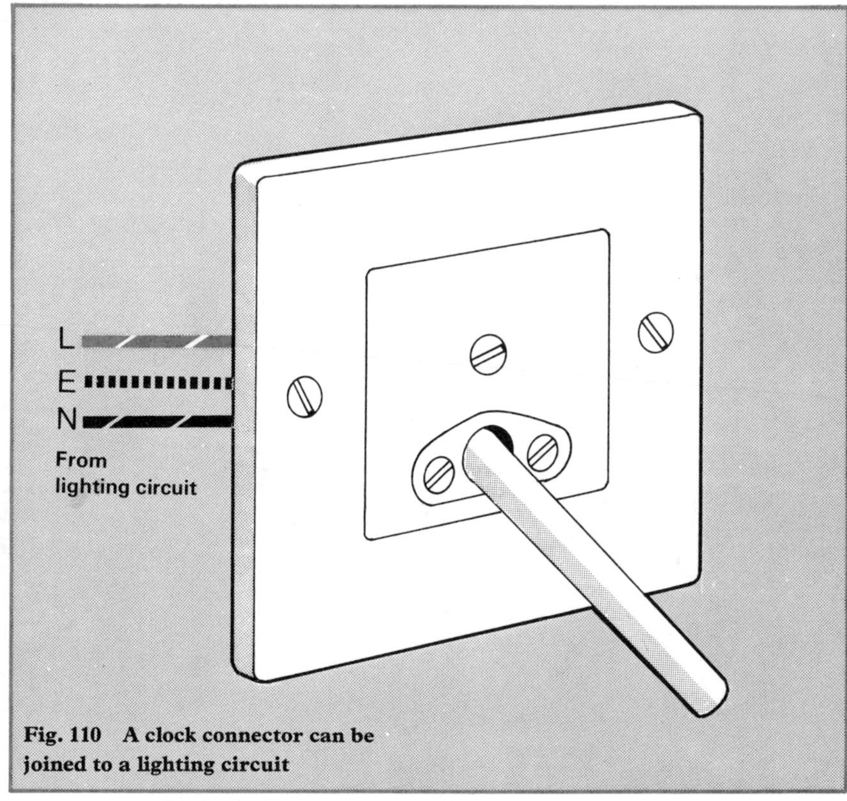

Fig. 110 A clock connector can be joined to a lighting circuit

IN THE BATHROOM

Fixed appliances fitted in a bathroom such as towel rails or oil-filled radiators should be supplied from fused connection units – not from socket outlets – which are wired from the ring circuit.

The connection units, which should contain pilot lights, are fitted near the appliances but out of reach of anyone who is in contact with the bath or shower.

CLOCKS OFF THE MAINS

Electric clocks run off the mains have to be supplied from a special socket outlet called a clock connector, *fig. 109.* These connectors comprise a fused plug and a socket outlet – in two sections.

The plug section is secured by a captive nut. This prevents it being accidentally removed from the outlet and thus stopping the clock.

The plug is fitted with a fuse which may have a 1, 2 or 3 amp rating, but is now normally 3 amp. The connectors are made in flush or surface types and you can also get a type which can be accommodated in a metal plaster-depth wall box such as that used to house a plateswitch.

An electric clock consumes very little current and can therefore be safely supplied by a lighting circuit. It is, however, also possible to use the ring circuit as the source of supply.

MORE THAN ONE CIRCUIT

If a number of clocks are to be supplied from the mains, it is better that they do not all run off the same circuit. If more than one circuit is used, not all the clocks will be out of action if a circuit fault develops. This could be awkward.

If you decide to utilise the lighting circuit, the cable to install is 1·0 $mm.^2$ two core and earth pvc sheathed. This cable can run from the live and neutral terminals of a lighting joint box or loop out from a ceiling rose to the socket outlet section of the clock connector, *fig. 110.*

In *fig. 110* an earth conductor is shown, but as old lighting installations are unlikely to have an earth conductor, it can, in this case, be disregarded.

Most electric clocks are, in fact, double insulated, so no earth connection is normally needed. However, some clock connectors are fitted with an earthing terminal.

If the cable to the outlet has an earthing conductor, the mounting wall box for the clock connector should be fitted with an earthing terminal to accommodate the conductor.

USE OF RING CIRCUIT

If you decide to use the ring circuit as the source of supply for an electric clock, you can loop into a 30 amp joint box on the ring and run from it a length of 2·5 mm.² two core and earth pvc sheathed cable to the connector, *fig. 111*.

This, however, could involve a long run of cable – an expensive item these days. A better idea, therefore, is to connect a *non*-switched fused connection unit fitted with a 3 amp fuse direct to the ring circuit. From the fused unit run a length of 1·0 mm.² two core and earth pvc sheathed cable to the clock connector.

The clock is connected to the plug section of the connector with two core circular sheathed flex. If the clock has a metal case and is not double insulated, three core flex will be needed as this has the necessary earthing conductor.

TELEVISION SETS

One appliance many people take for granted in the home is the television set. Often it is not treated with the respect it deserves.

There is, of course, very little you can do if the set should go wrong apart from calling in a qualified service engineer. In fact, television sets should never be interfered with by anyone who is not qualified to service them. They are complicated pieces of equipment carrying very high voltages and are therefore potentially dangerous.

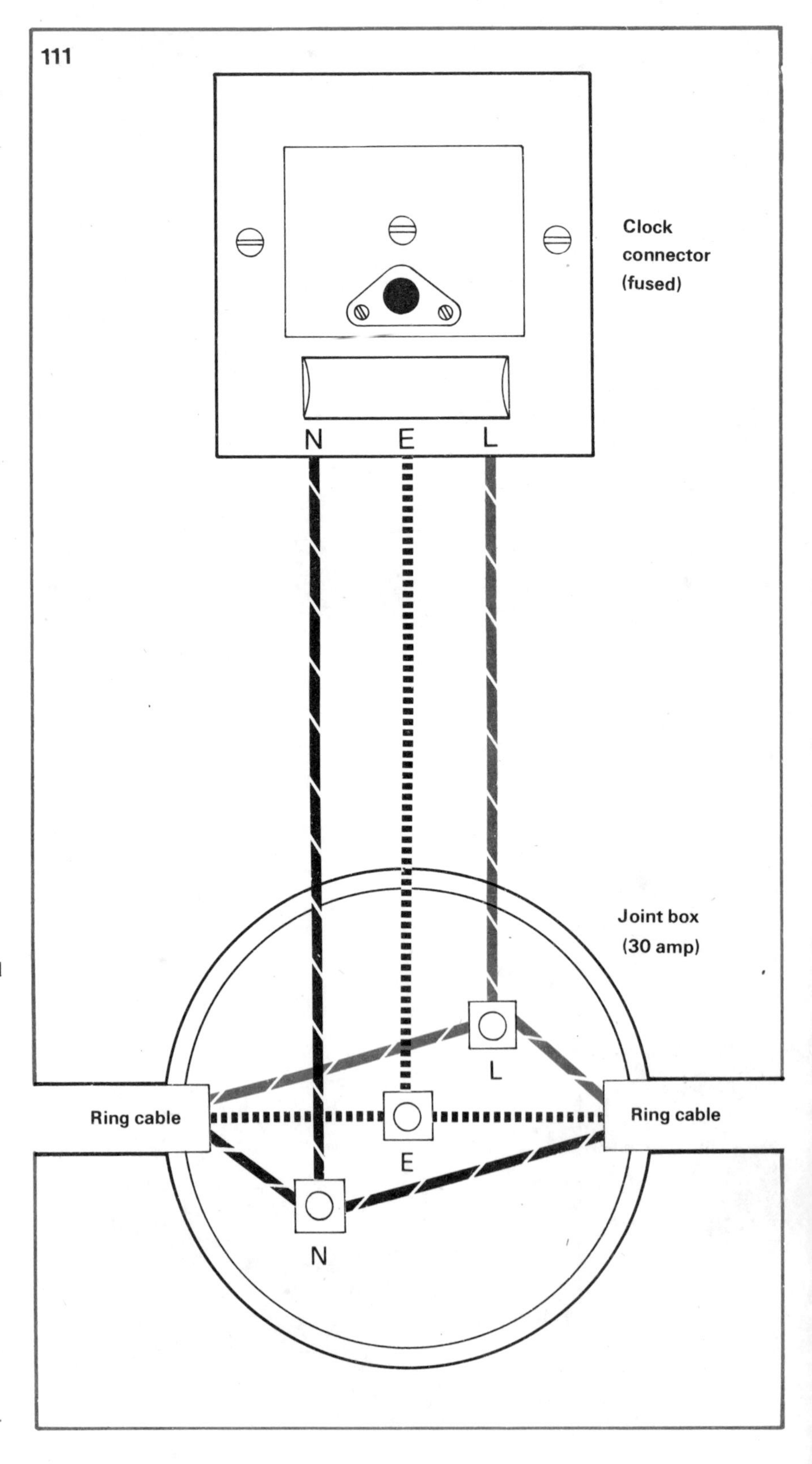

What you *can* do is to take precautions when using the set. First and foremost, the socket outlet on the wall to which the set is plugged in, must be correctly wired up. If it isn't this could cause a fire.

In most cases, of course, the outlet will be properly wired and safe to use. If, however, your installation is at all suspect, have it attended to before connecting a TV set (or any other appliance for that matter) to it.

If you have to install a new socket outlet on the ring circuit to supply a TV set, make sure you have got all the conductors connected up properly. I cannot repeat this too often: the red conductors go to the live (L) terminals; the black to the neutral (N) terminals; and the earth (green) conductors to the earthing (E) terminals. I repeat these connections in *fig. 112.*

Although this may seem obvious, it is vitally important. If, for example, the black and red conductors were reversed in the outlet terminals, parts of the TV set and the flexible cord leading from it would become permanently *live*. This error could cause a fire even if the switch (if any) at the socket outlet and the switch on the set itself were turned off.

This is because the reversing of the cables would mean that the plug fuse and the outlet switch would both be in the neutral lead and not in the live lead as they should be. Switches should always be connected in the live lead.

To be on the safe side, it is better not to rely on the switches but to pull out the plug from its outlet whenever the set is not in use. This is the most efficient way of ensuring that the set or other appliance is completely isolated from the mains.

Whatever you do, it is in any case a good plan to use a switched socket outlet for a TV set. If you have to use an unswitched outlet, however, you can get a switched plug fitted with a pilot light. This at least will show you whether the set is off or on. But for complete isolation of the set, you will still have to pull out the plug.

Fig. 111 Ring circuit connections for clock connector

Fig. 112 Socket outlet wiring on the ring circuit for a TV set

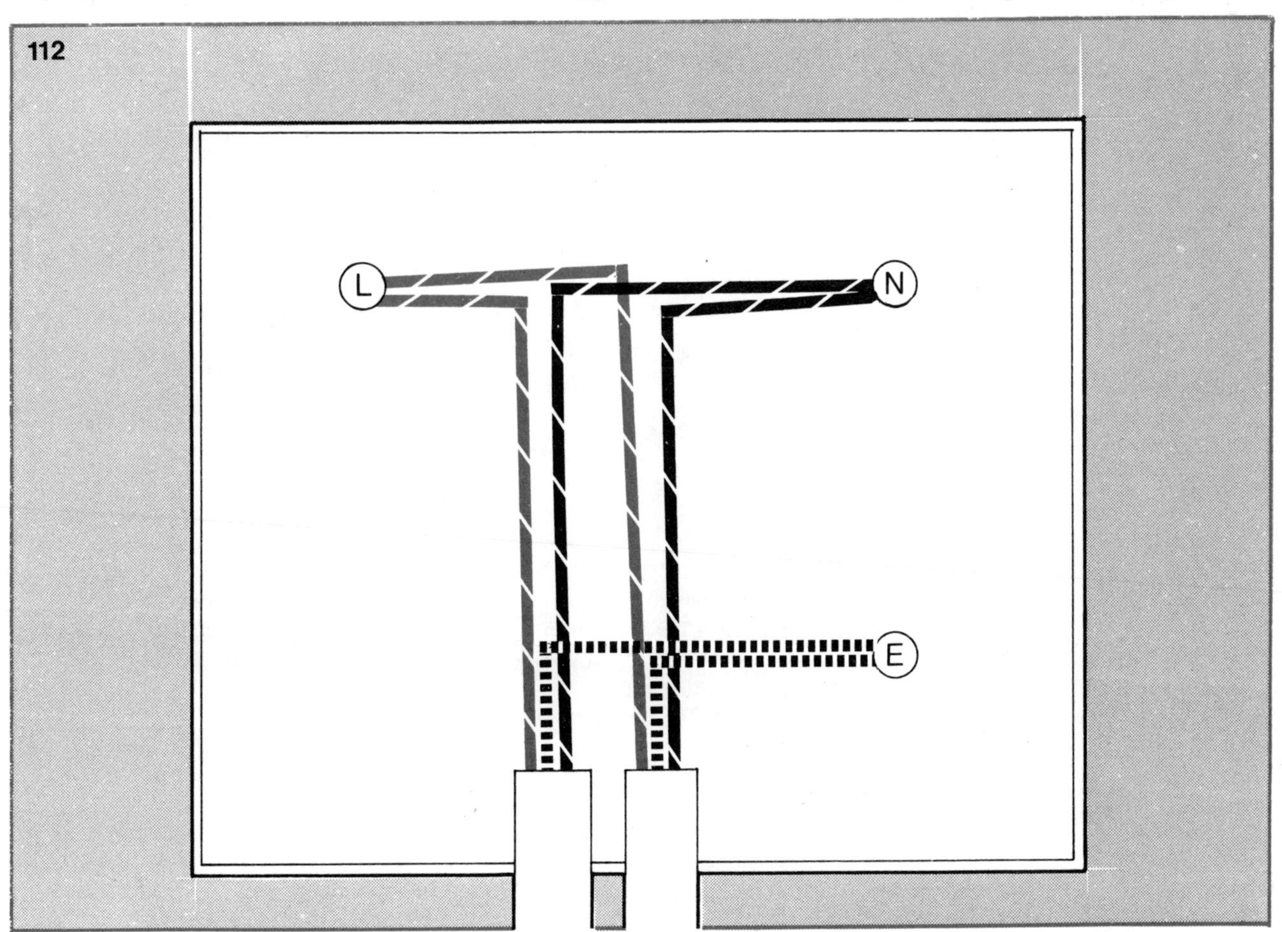

EARTHING

Like many electrical appliances, the majority of TV sets have no earth connection and are fitted with a two core flexible lead, the earth pin of the three-pin plug being left unoccupied.

Never connect your set to a plug which has only two pins, nor to the type of plug which is sometimes used in a lampholder. This could cause the chassis of the set to become live and therefore dangerous if the back of the set were removed while the current was switched on.

Also, if the control knobs happened to become damaged and any metal part of them became exposed to the touch, this part, too, could become live.

ELECTRIC BLANKETS

Electrical appliances which have been responsible for a large number of fires and accidents in the home are electric blankets.

Nowadays, these appliances conform to a very high safety standard, and many of the fires caused are due to misuse by the owner of the blanket. It is therefore essential that the instructions supplied by the manufacturer should be followed to the letter.

Never attempt repairs to an electric blanket. Get it serviced regularly by experts and leave it with them to put right if the appliance goes wrong. Care of the blanket will cut repairs to a minimum.

Safety measures you should take when using electric blankets include:

1: Keep the switch well out of reach of children.

2: Under blankets should not be kept switched on when a person is in the bed.

3: Don't fold blankets up when not in use as this can bend and possibly fracture the heating cable.

4: Keep the instructions safely with the blanket so that anyone unfamiliar with it can read them before use.

5: Avoid buying a blanket which is not made by an established and recognised company.

HEAT/LIGHT UNITS

There are a number of electrical appliances on the market today which supply both heat and light and are known generally as heat-light units. You often find them fitted in bathrooms and kitchens in place of the usual ceiling light, but there are types which are designed to be used in other rooms.

Many of them have a heating unit with a loading of 750 watts and a light rated at 100 watts – a total of 850 watts.

These units are often connected to the lighting circuit, but they should not be. The reason is that there is a risk of overloading such a circuit with so highly rated an appliance.

A lighting circuit is rated at 5 amps and its maximum load should not exceed 1,200 watts. This is the equivalent of twelve 100 watt lamps.

As I said in chapter 10, each lamp rated up to 100 watts is, for the purpose of specifying the number of lamps for a circuit, assessed at 100 watts regardless of whether its actual wattage is 100 or below. Every lamp over 100 watts, however, is assessed at its true rating.

Therefore, if you fit a heat/light unit of 850 watts, you leave a margin of only 350 watts in reserve for other lights on the circuit.

OVERLOADING

Let's see what could happen in a typical family residence, with only one lighting circuit, one winter's evening where a heat/light unit of this size is fitted in the bathroom.

Mother decides to take a bath and switches on the unit: watts in use, 850.

Daughter sits reading in the lounge under a standard lamp fitted with a 100 watt bulb. Total watts in use, now 950.

Two wall lights burn also in the lounge. Although these may be only 60 watt lamps, they are assessed at 100 each. Watts now in use, 1,150.

So far so good, but supposing father decides to go into the kitchen and switches on the fluorescent light? If this has a 100 watt tube, it will be assessed at 200 watts. So the total of watts rises to 1,350 and the circuit then becomes overloaded, in theory anyway.

At this stage, the circuit *may* not, in practice, be overloaded because if we had used the true rating of the two wall lights and the fluorescent tube, the actual wattage being consumed would be 1,170 – 30 watts below the permitted 1,200 which it the total for the circuit.

MARGIN TOO SMALL

However, this margin of only 30 watts is not sufficient to take into account the possibility of one more light being switched on, say in the toilet or in a bedroom, even momentarily. If that happened, then the circuit would definitely be overloaded and the circuit fuse would blow.

This can be disturbing at the best of times, especially if all the lights are on the same circuit and therefore all fail simultaneously. It could be dangerous to anyone taking a bath, for example, and even more dangerous for anyone using a power tool supplied off the ring circuit.

Although the fusing of the lights would not affect the working of the power tool, the shock of being plunged into sudden darkness could cause a nasty accident to the person using the tool.

No doubt many of these heat/light units have been used off a lighting circuit, but it is a practice to be deplored. One person living alone (perhaps even two) might remember to keep other lights turned off while the unit is in use, but it is asking too much to expect a family to do so at all times.

Therefore it is better to play safe and avoid fitting such a unit of so high a wattage in place of an ordinary lamp.

There are, fortunately, alternative methods of wiring such a unit. It can be supplied either from a separate 5 amp circuit run from the consumer

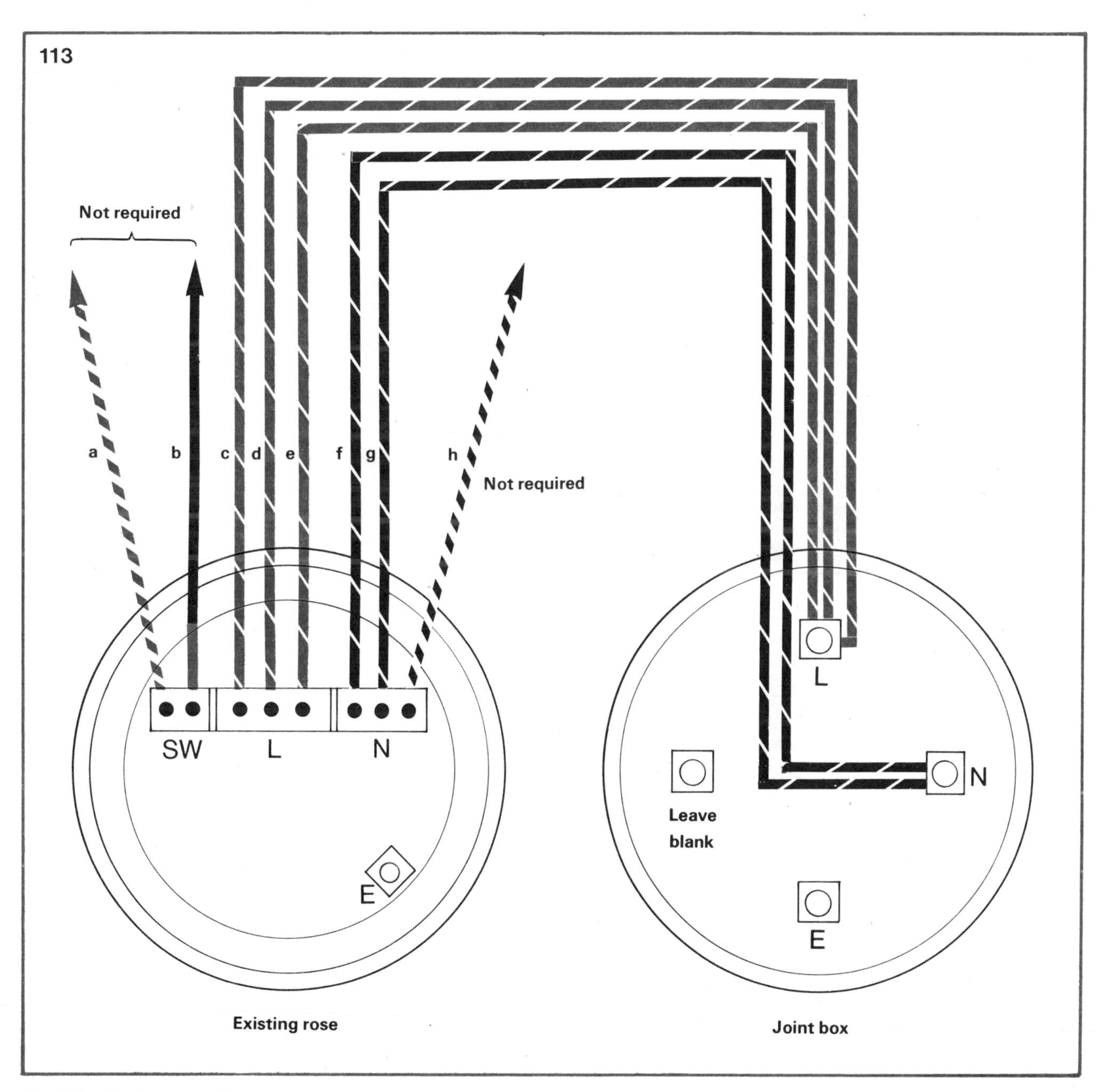

Fig. 113 Sealing off ceiling rose wires in a joint box
a) Flex live wire
b) Switch wire
c) Live feed
d) Live switch wire
e) Live loop to next rose
f) Neutral feed
g) Neutral feed to next rose
h) Flex neutral wire (not required)

unit, or from a fused connection unit joined to the ring circuit.

If you plan to fit the unit in place of an existing light, the cables at the ceiling point, will no longer be needed as they will be connected to the lighting circuit. They will have to be disconnected from the ceiling rose and terminated (sealed off) at a joint box fixed at a point above the ceiling, on the joists, *fig. 113*.

DISCONNECTING

While doing this, great care will have to be taken to be absolutely sure that the connections are not accidentally changed. There may well be a confusion of cables at the lighting point. It is impossible to say exactly what the wiring arrangements will look

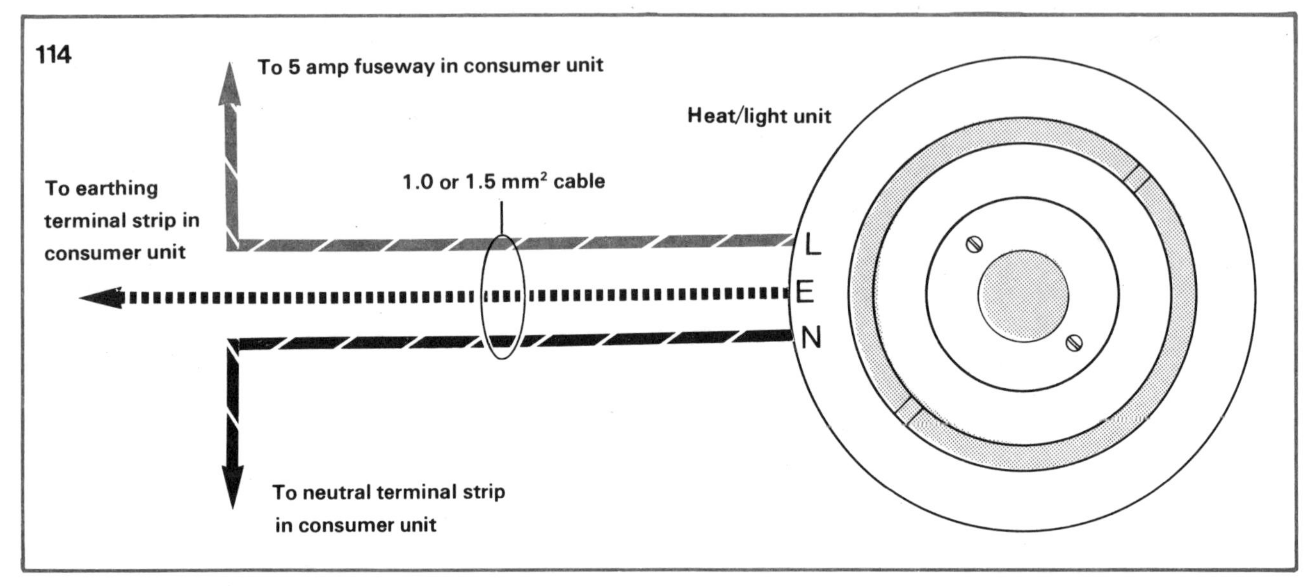
114
To 5 amp fuseway in consumer unit
Heat/light unit
To earthing terminal strip in consumer unit
1.0 or 1.5 mm² cable
L
E
N
To neutral terminal strip in consumer unit

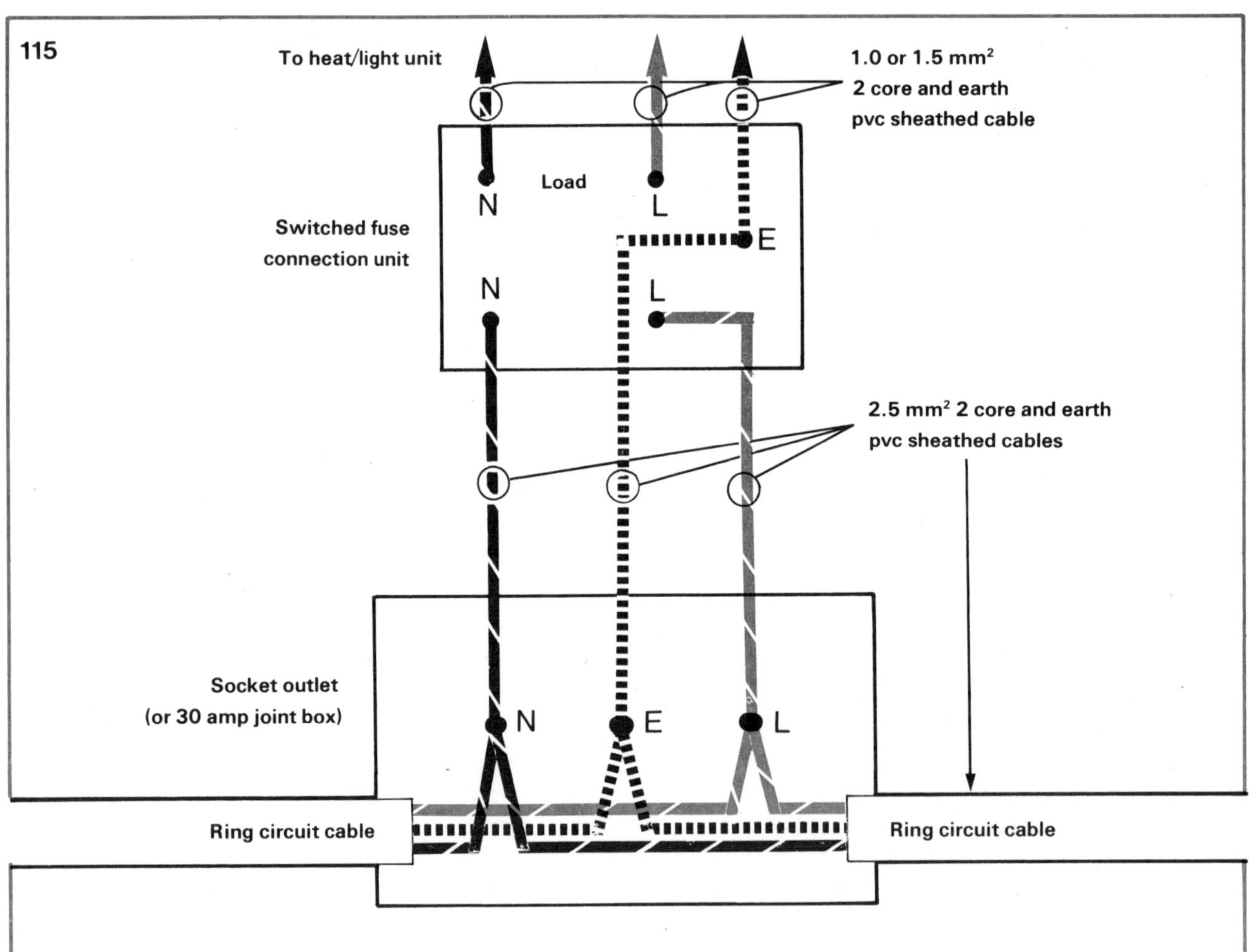
115
To heat/light unit
1.0 or 1.5 mm² 2 core and earth pvc sheathed cable
Load
N
L
Switched fuse connection unit
E
N
L
2.5 mm² 2 core and earth pvc sheathed cables
Socket outlet (or 30 amp joint box)
N
E
L
Ring circuit cable
Ring circuit cable

Fig. 114 Heat/light unit supplied from consumer unit

Fig. 115 Heat/light unit supplied from ring circuit

like, so if you cannot understand the cable layout, postpone the job until you can get expert advice. And do remember to cut off the current at the main switch before you remove the rose cover in order to make your initial inspection.

One way to avoid mixing up the cables, once they have been identified, is to disconnect one conductor at a time and label it accordingly. There will almost certainly be more than one conductor at the live and neutral terminals. Make sure that these are kept together on the same terminal when they are eventually transferred to your joint box.

The earthing terminal in *fig. 113* on the old rose has been purposely left blank, as has the one in the joint box. The reason for this is that old installations are unlikely to have earthing conductors at lighting points.

If, of course, your installation has an earthing conductor at this point, it should be transferred to the joint box earthing terminal.

NOT REQUIRED

The old switch wires and the flex wires at the lighting point will not be required for the heat/light fitting and should be disconnected. Be very careful when looking for the switch wire not to remove the wrong wire.

The switch return wire, as it is called, will probably be one conductor (the black one) of a two-core cable, the red conductor of this cable going to the live terminal on the rose. But although it is black, remember that it will, in fact, be live!

As I said in chapter 11, the switch wire should carry a short length of red insulation sleeving to identify it (at the switch terminal), but rarely is this done.

Figure 114 illustrates how the unit can be connected up if you decide to run a new circuit from the consumer unit.

Figure 115 shows how to join the unit up to a ring circuit using a fused connection unit. This unit should be fitted with a 13 amp fuse, unless the instructions specify otherwise.

Wiring Outside

If you want a lighting point or socket outlets in the garden shed, garage or greenhouse, special equipment must be used to meet the wiring regulations.

A special type of cable is run from the house but this should be separate from the house circuits. This cable can run underground, but it must be buried at least 18 inches below ground level, *fig. 116*. Alternatively, the cable can run overhead or along a wall. *It must not run along a fence.*

The cable has to be controlled by its own main switch. This is fitted near the electric meter in the house.

The installation in a shed (or whatever) has to be controlled by an isolating switch which is fixed in the shed.

This switch can cut off current at the shed, thus isolating the circuit and the appliances from the mains.

Fig. 116 Outside cables can run under the ground

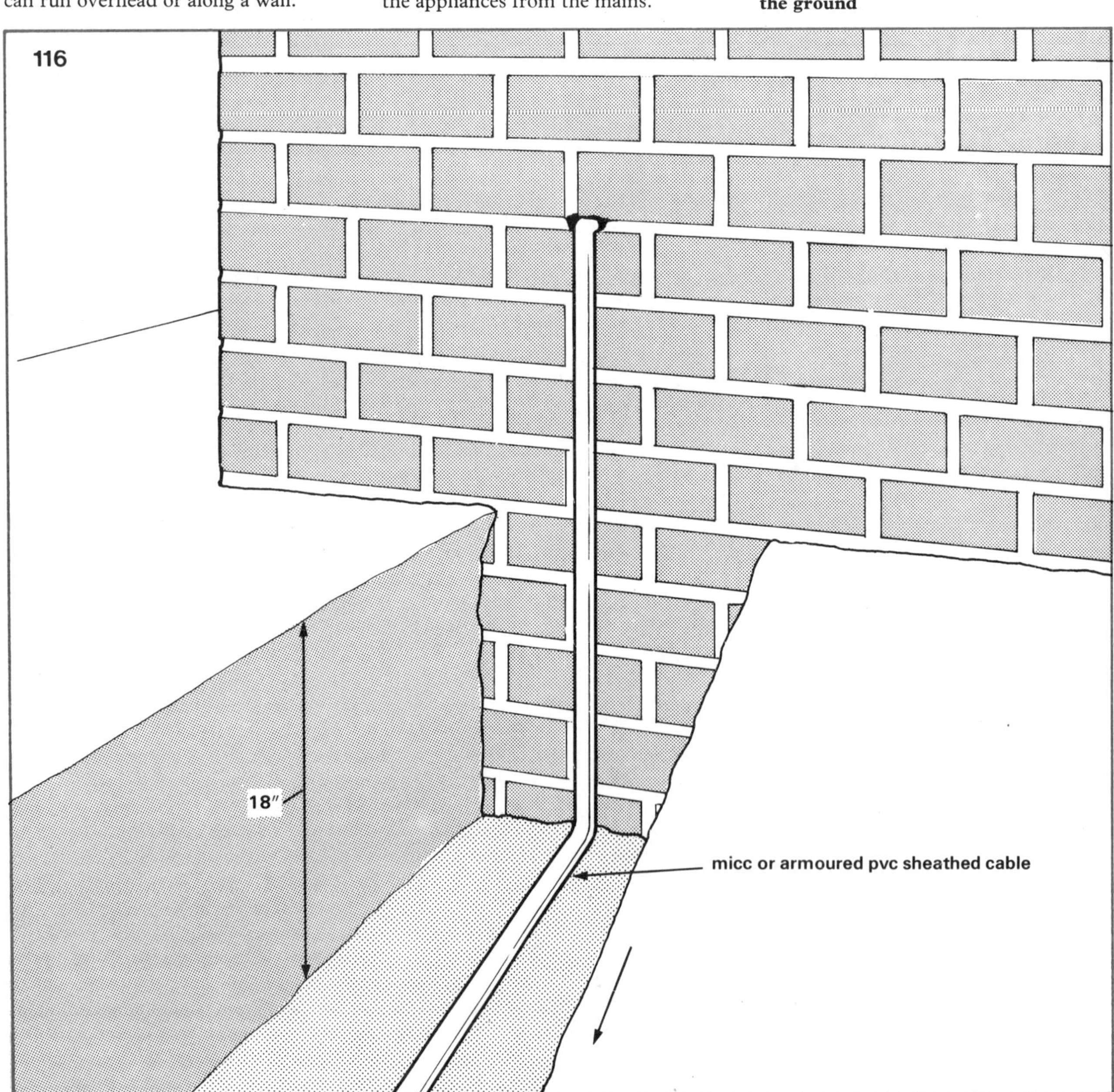

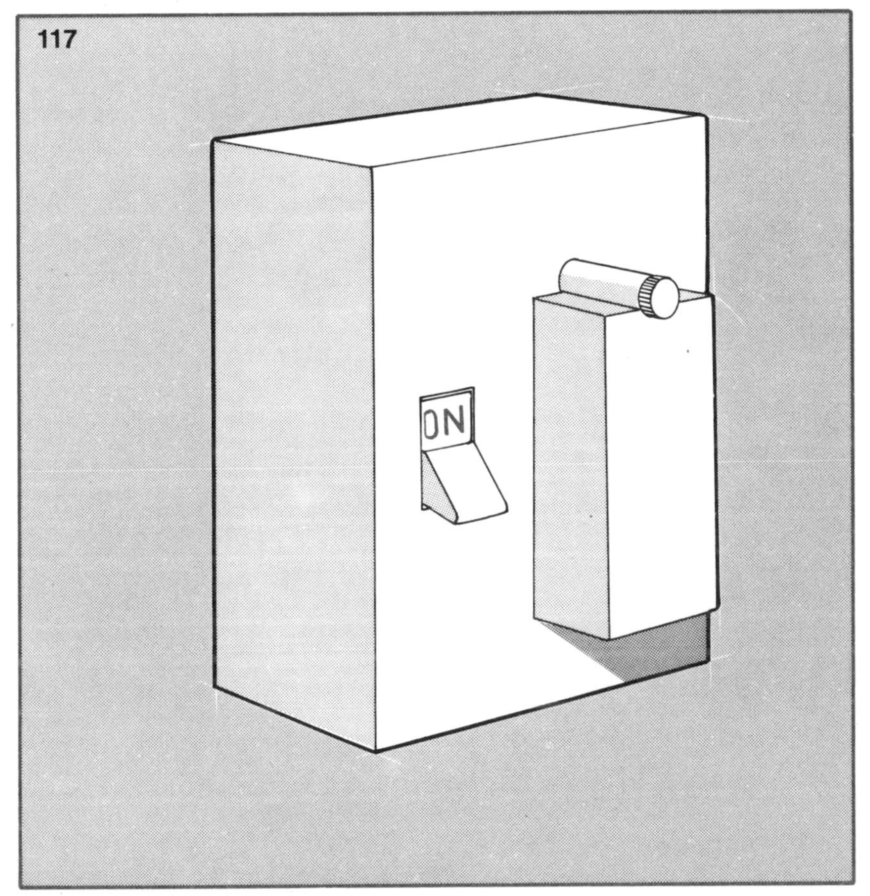

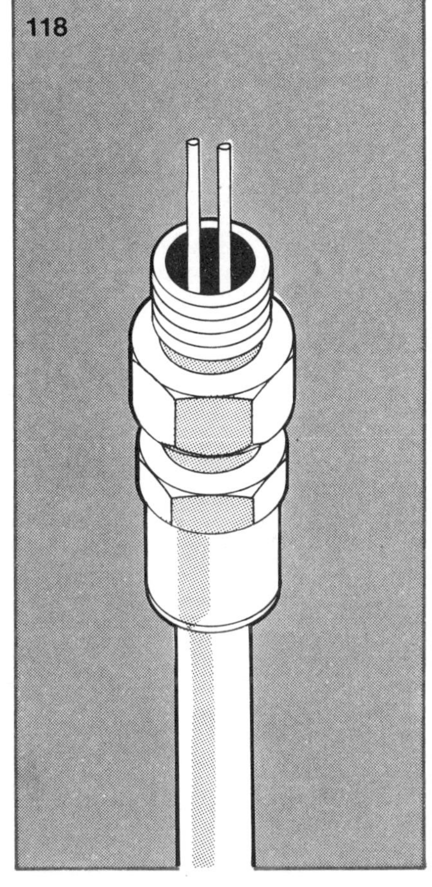

Fig. 117 Switch/fuse unit (MEM)

Fig. 118 Ends of outside cables terminate in screwed glands

You have a choice of main switches: a switch-fuse unit (a combined main switch and fuse), *fig. 117*, or the main switch of a separate consumer unit.

A switch-fuse unit can supply only one circuit for lighting and power, but is enough for modest needs. A consumer unit, of course, can supply several circuits, according to its size.

CHOICE OF CABLE

Two types of cable can be used underground (a) mineral insulated copper clad (m.i.c.c.) or (b) pvc armoured pvc sheathed. See separate chapter on cables.

These cables have only two cores – live and neutral – but no earth wire.

The copper sheath of (a) and the armoured core of (b) themselves act as the earthing conductors. If you order m.i.c.c. cable, stipulate that it must have a pvc covering.

For most purposes, the 2·5 mm.² size cable will be adequate. If, however, the load demand is above 4·8 kW (4,800 watts), or the cable is longer than 70 ft., use 4·0 mm.² cable.

You can lay these cables in a trench without giving them any other protection, but if you indulge in double digging or trenching in the garden, bury them deeper than 18 inches.

TERMINATIONS

Holes will have to be cut in the walls of the house and of the shed for the cables to pass through. Whether you use pvc armoured pvc or m.i.c.c. cable, its ends should terminate in screwed glands, *fig. 106*. You can get the cable with these already fitted. If you use m.i.c.c. cable, this will need glands *and* seals. The seals can be fitted at the shop.

To terminate the cable at the house end, you need a metal knockout box of the same type as that used to house a flush socket outlet. Remove a suitable knockout and pass the house end of the cable, with glands fitted, into it. This box is fitted just inside the house.

Screw a conduit locknut on to the end of the gland. This ensures good earthing continuity. From the knockout box run a cable of the same size, but twin and earth pvc sheathed, to your main switch in the house, *fig. 119*.

The joints in the metal box (between this cable and the outside cable) can be made using a special terminal block, *fig. 120*. After making the joint, fit a blanking plate, *fig. 121*, over the box.

Make sure, by the way, that this box has an earthing terminal. From that terminal run a short green insulating wire to the earthing terminal of the terminal block in the box, *fig. 120*.

Fig. 119 Cables joined at terminal block in wall box

Fig. 120 MK terminal block

Fig. 121 MK blanking plate

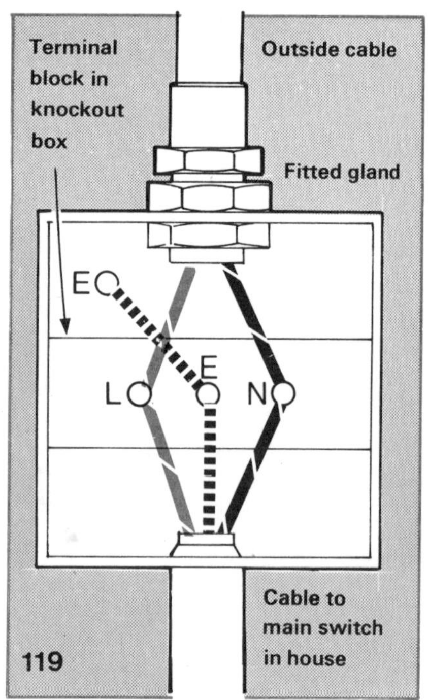

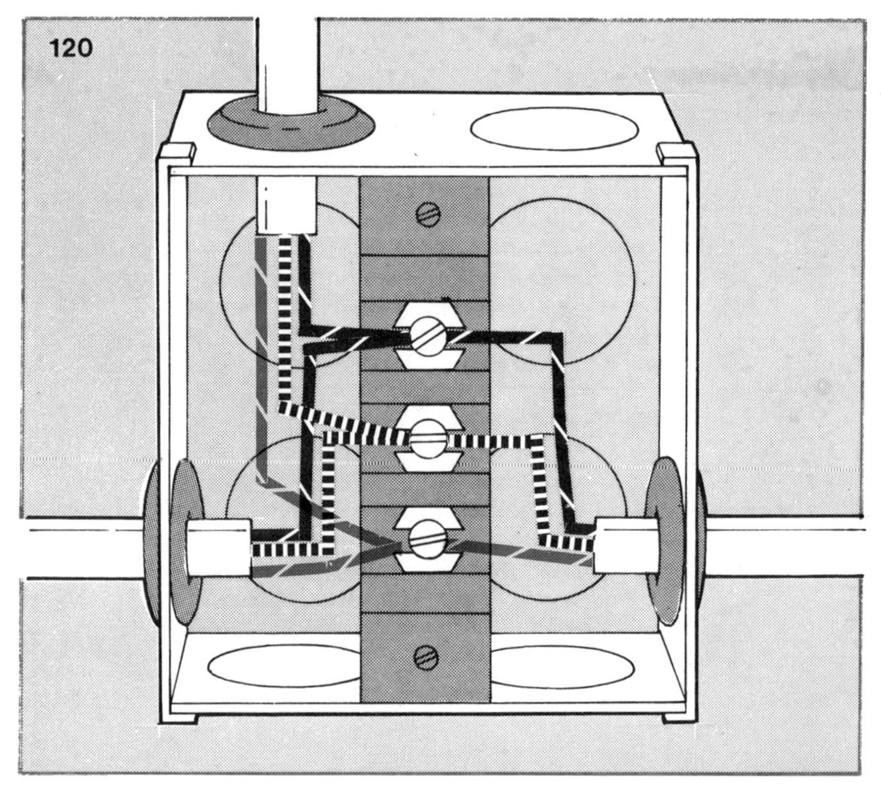

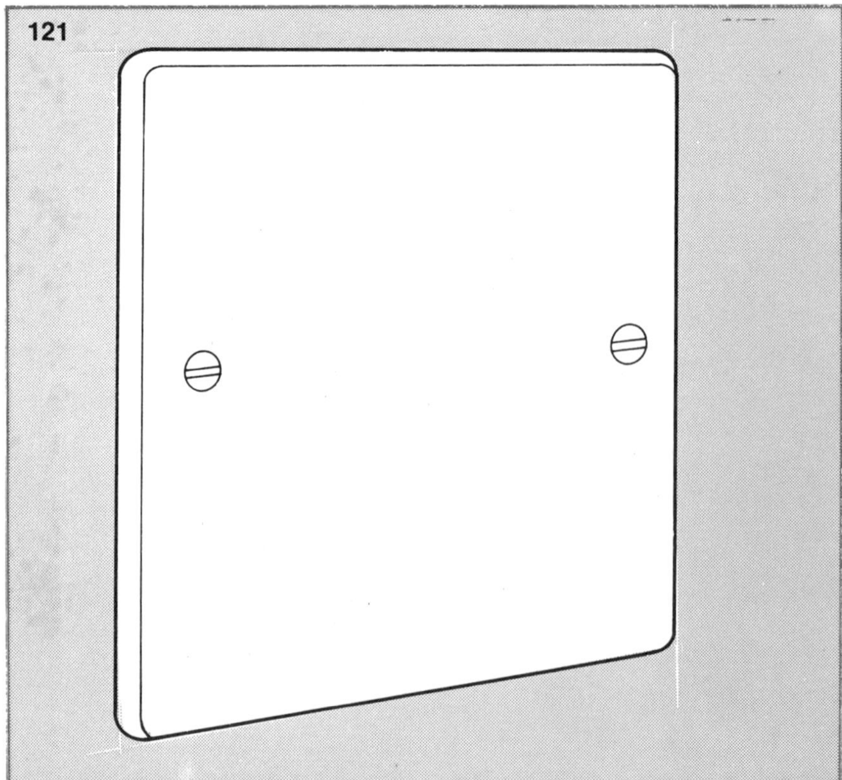

The other end of the outside cable should run direct into the isolating main switch in the shed. So fix the isolating switch near the spot where the cable enters the shed.

If this is not possible, use a metal knockout box and the same method of jointing as for the house end of the cable.

ALTERNATIVE

For a shed, you will probably need only a light and perhaps a couple of socket outlets. If so, you can forget about a main switch at the shed end. Instead, you can use a 20 amp double pole switch (plateswitch type) and a fused connection unit fitted with a 13 amp fuse.

These can be mounted on a dual pattress box and linked with a short length of 2·5 mm.² twin and earth pvc sheathed cable. The connections are shown in *fig. 122.*

Fig. 122 Double pole switch and fused connection unit for power and lighting in outbuilding

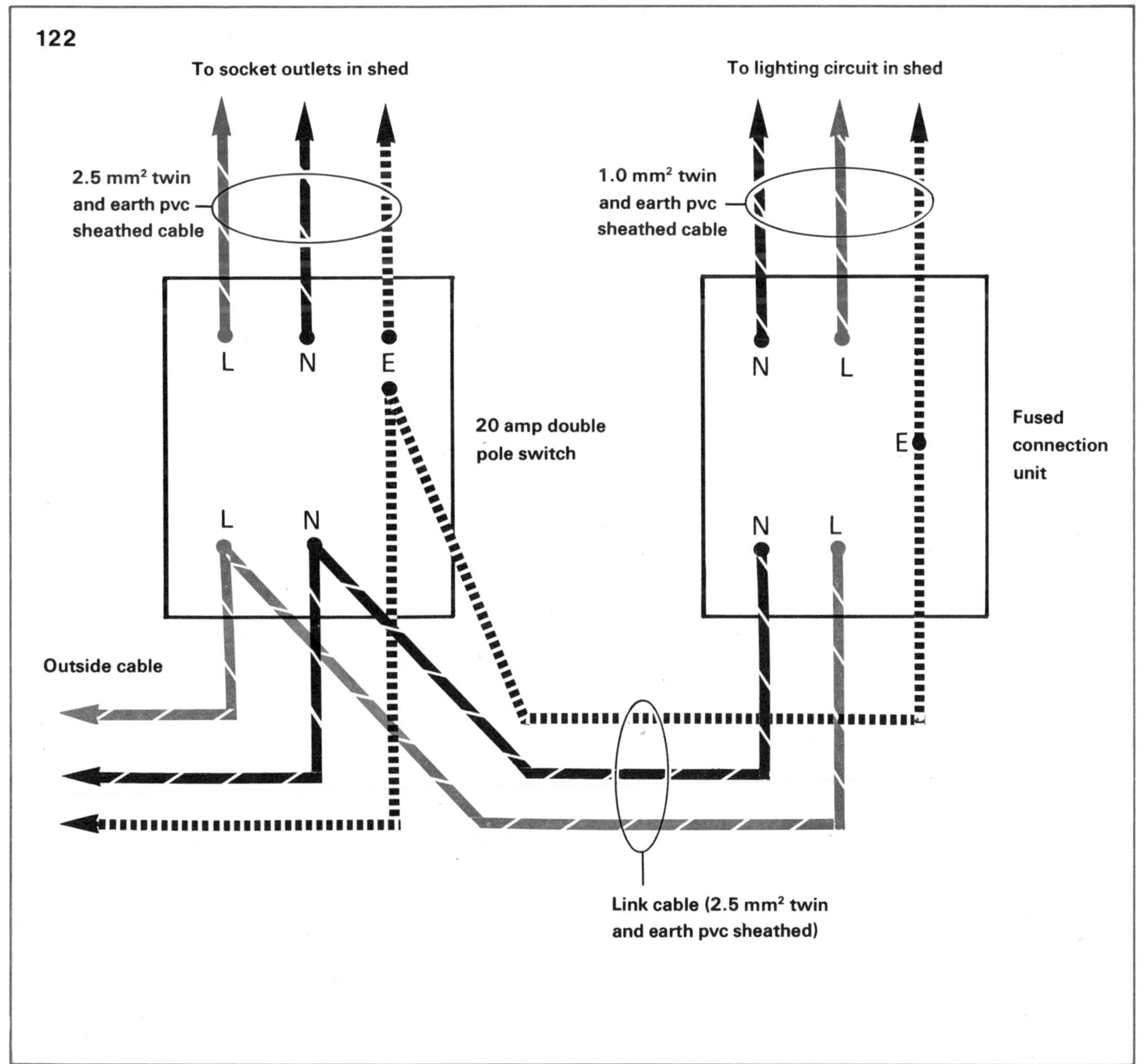

OVERHEAD WIRING

If it is not possible to bury the cable underground, you can use the overhead method of wiring, *fig. 123*. The cable required in this case is 2·5 mm.² twin and earth pvc sheathed, as used in house wiring. But if your power requirements exceed 4·8 kilowatts, or 20 amps, use a larger cable.

Provided the shed to be wired is situated not more than 10 ft. away from the house, the cable will not need any extra support. The cable should be fixed at least 12 ft. from the ground.

If the distance between the shed and the house is greater than 10 ft., you should hang the cable from a suspension wire. This is called a catenary and is strung between the two buildings.

For this you need a length of galvanized stranded wire and eye bolts for fixing it.

The cable should be fixed to the suspension wire at intervals of about

Fig. 123 Overhead wiring method for outside lighting or power

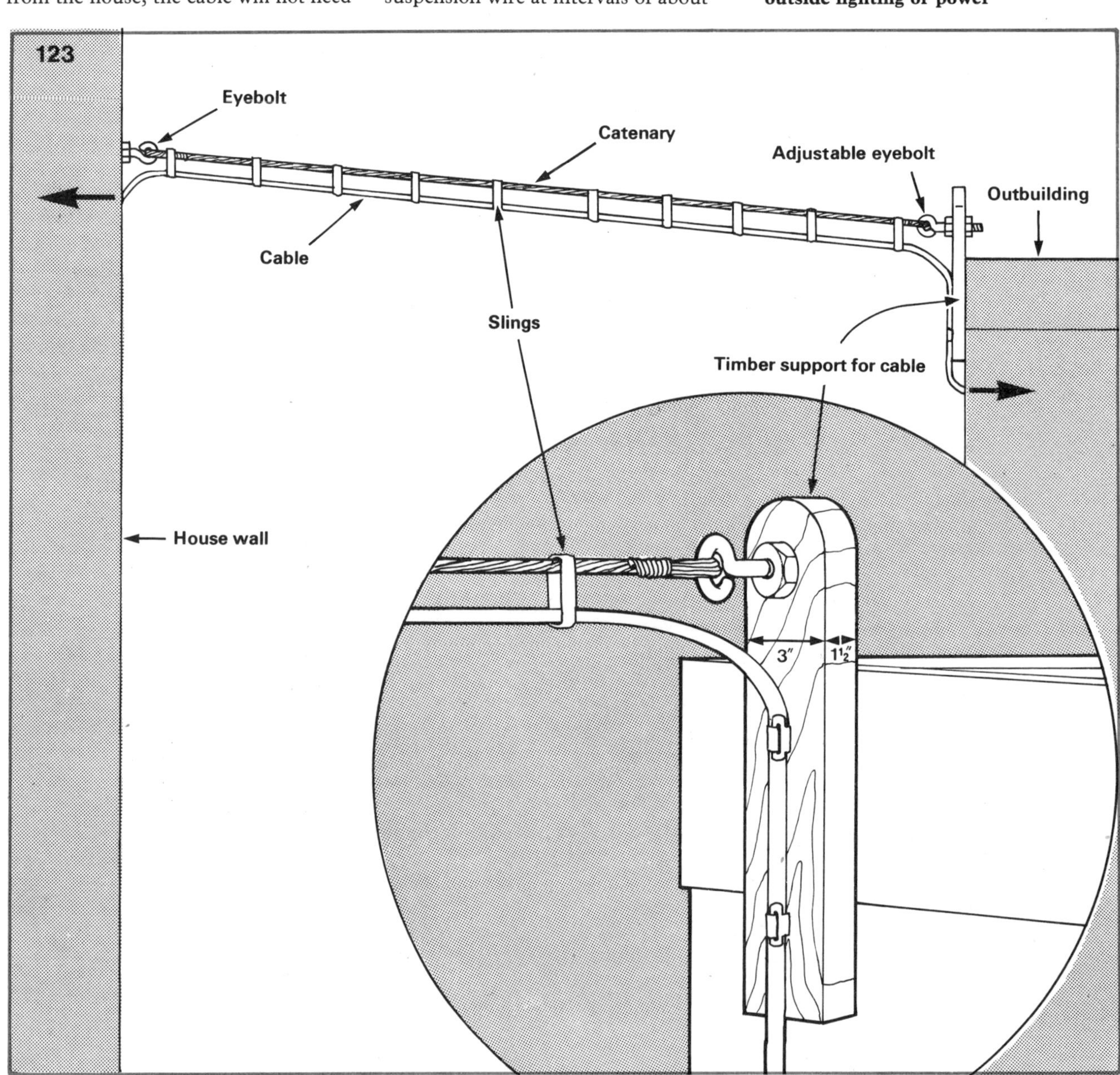

9 in. You can do this by using plastic slings available from electrical accessory shops.

Fix the suspension wire at one end with an adjustable eye bolt and a straining eye bolt at the other. The suspension wire should be connected to earth (the earthing terminal of the main isolating switch in the shed.)

In no circumstances, however, can the suspension wire be used as the earth continuity conductor for the outdoor circuit. The conductor which should be used for this purpose, of course, is the bare earth wire in the cable itself.

USING CONDUIT

Another method of wiring is to run the overhead cable in conduit. This should be a continuous length of heavy gauge steel. The conduit should be long enough to enter both the house wall and the wall of the shed. If you

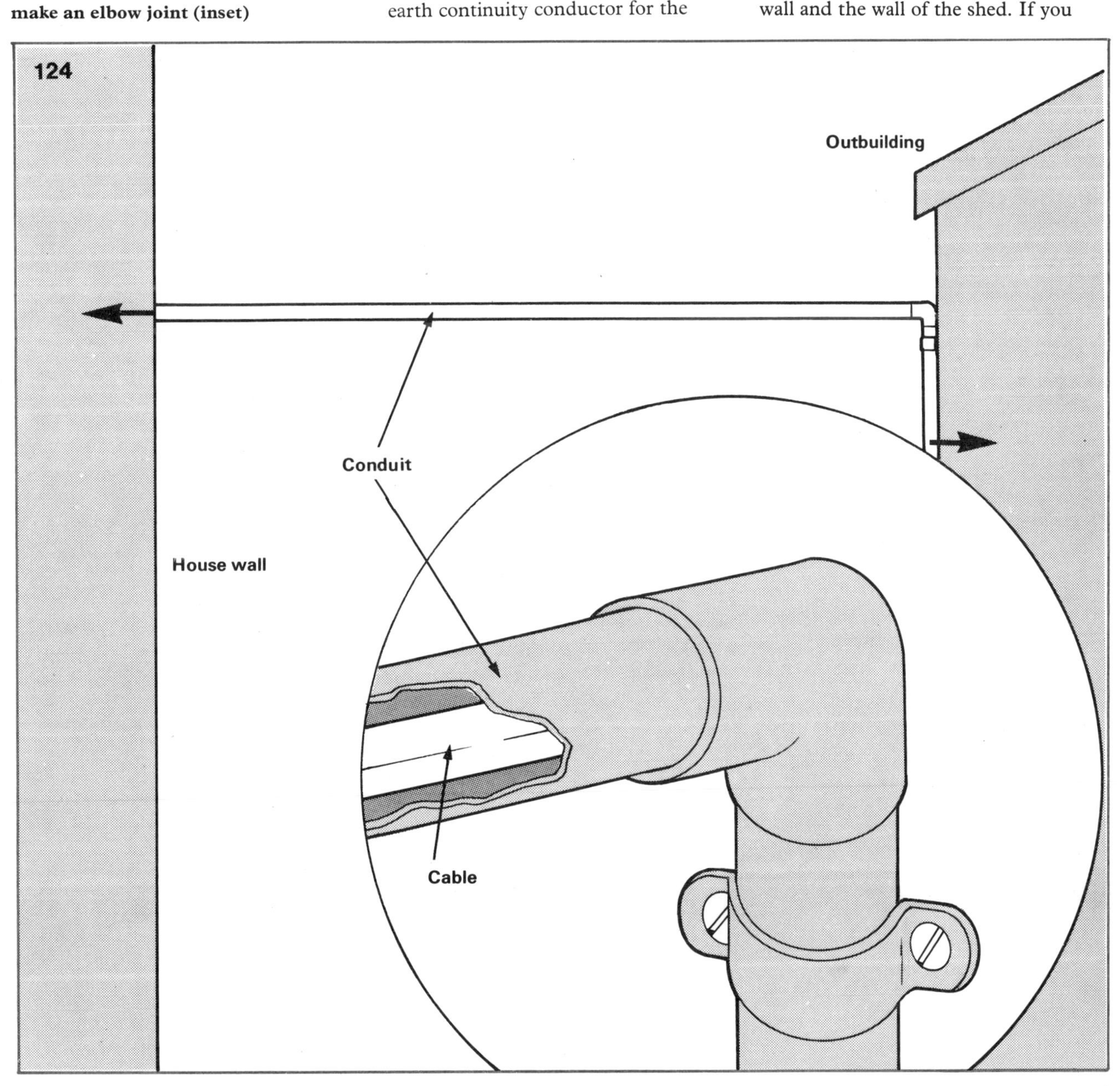

Fig. 124 Using conduit for overhead cable. It is best if the conduit goes through the wall. If it has to run down, make an elbow joint (inset)

use metal conduit, this must be connected to earth (bonded, as it is called).

To do this, fix an earthing clip, *fig. 125*, on one end of the conduit. To the clip connect a green insulating cable. This runs to the earthing terminal of your isolating switch in the shed.

At the house end, connect two lengths (about one metre each) of 6·0 mm.² single core pvc sheathed cable to the switch/fuse unit or main switch of your new, separate consumer unit. Connect the red conductor to the live terminal and the black to the neutral.

Leave these wires for the electricity board to connect to the mains. They will do this after you have filled in an application form and arranged with them a date for the circuit to be tested, approved (we hope) and connected.

If you are at all doubtful about your ability to carry out outside wiring (or, indeed, any form of house wiring), get expert advice.

As I have emphasised, electricity is potentially dangerous, especially if mishandled, so make sure you know exactly what you are undertaking *before* you start.

Many of the accessories and cables mentioned in these pages can be obtained from department stores and electrical shops. For some materials, however, you may have to go to an electrical contractor.

Whatever you do, always buy materials made by a reputable firm. Don't be tempted by cut price items. They may be adequate for a time, but you cannot be certain.

By doing your own electrical jobs you are saving money, so you can afford to buy the best available.

Fig. 125 Earthing clip

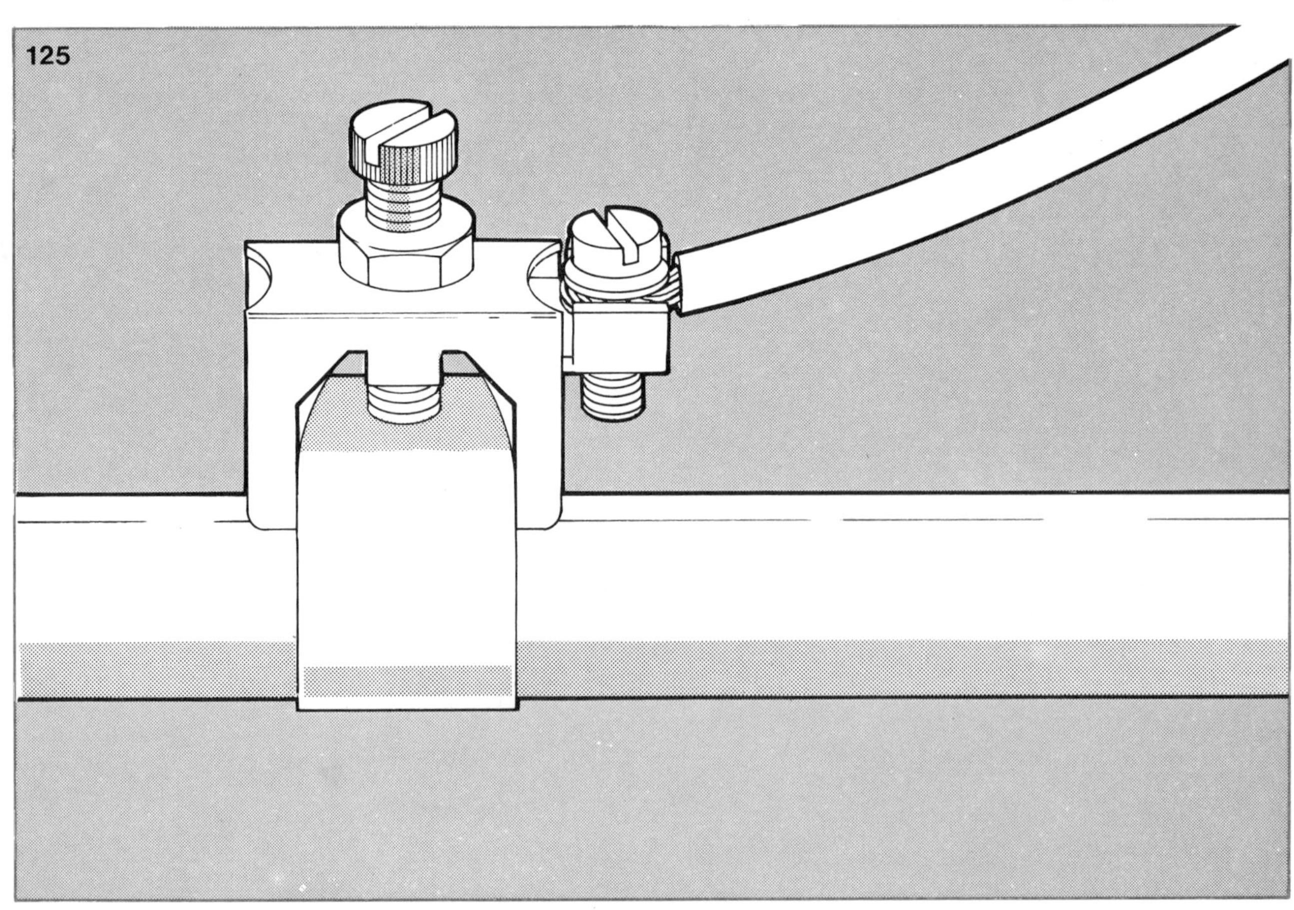

Questions and Answers

A lot of work involved in wiring is repetitive. In these typical questions wiring details which appear elsewhere in the book have, in some cases, been omitted from the answers.

Q. *Can pvc sheathed cable be dropped down the cavity of a wall or must it be enclosed in conduit?*

A. There is no need to run this type of cable in conduit.

Q. *Can I run a five kilowatt water heater off a 30 amp ring circuit?*

A. No. Nothing higher than a three kilowatt outlet can be run off a socket outlet on a ring circuit. You need a separate 30 amp circuit for this type of heater. It should be supplied by a 4 mm.² cable.

Q. *Is it possible to stop clicking noises in a television loudspeaker? These can be heard whenever a light switch is operated.*

A. The only way to stop the clicking noises is to fit a suppressor on every switch – an expensive cure! The clicking is harmless and nothing to worry about.

Q. *Can an electric lawn mower be plugged into a two-pin 15 amp socket outlet?*

A. Provided the mower has a two-core (not a two-core and earth) flex, this is permissible, but the lawn mower must be a double insulated type. If it is, it will not require earthing.

It would be better, however, to fit a 15/5 amp adaptor to the plug and fit the 5 amp section of it with a 5 amp fuse. The fuse to protect the mower should not exceed 5 amps.

Q. *Can we raise our socket outlets to about 2ft. above ground level? At present they are fixed just above the skirting board. We have solid floors and live in a bungalow and the wires are all run in conduit.*

A. As the cables will not be long enough to reach the new outlet positions, you cannot do this without blanking off all the old sockets. However, if you don't object to a lot of blanking plates in the old positions, it can be done.

Buy a blanking plate for each outlet and a similar number of three-way terminal blocks and steel wall boxes. Turn off the current at the consumer unit, disconnect each socket outlet and connect the cable conductors to the terminal block. Make your new holes for the outlets and fit the new wall knockout boxes.

From the terminals of each terminal block run a short length of 2·5 mm.² two core and earth pvc sheathed cable to the socket outlet terminals. Fit the outlets in their new boxes.

You will, of course, have to channel out grooves in the plaster for the new lengths of cable. Fit the blanking plates over the old wall boxes to hide the jointed cables.

The best time to do this job is before decorating – a room at a time.

Q. *Is it possible to get fluorescent light fittings which do not hum?*

A. Ask for a lamp ballast type of fitting. These are fitted with a filament lamp instead of a choke.

Q. *What size of conduit is needed to accommodate a 4 mm.² two-core and earth pvc sheathed cable to run underground to a garden shed?*

A. To cope with the bends necessary at each end of the cable, you will need 1¼ in. diameter conduit. If, however, you use armoured or m.i.c.c. cable, conduit will not be needed.

Q. *Can a flexible cord be sunk in plaster?*

A. No, not unless the flex is contained in conduit for the whole of its length.

Q. *What is meant by voltage drop when referring to cable sizes?*

A. As the term suggests, voltage drop represents the amount in volts by which the voltage of a circuit falls at the end of a cable run. This depends on the length of cable run and on the current.

All cables have a current rating with a voltage drop per ampere for each metre of the cable run.

The regulations lay down that the amount of voltage drop must not be higher than 2½% of the mains voltage at any part of a circuit. For example, with a voltage of 240, the 2½% maximum represents six volts.

So if a cable runs at its stated current rating, the length it is allowed to run must be such that the voltage drop is no more than six volts.

The length a cable can run before six volts are exceeded depends on its size and current rating. For instance, a 2·5 mm.² cable with a current rating of 21 amps can run for 56 ft. before its maximum is reached. A 6·0 mm.² cable has a maximum run of 82 ft.

Q. *Are all domestic appliances double insulated?*

A. No; only those which have no earthing conductor in their flexible lead. Appliances which are classed as double insulated have a separate means of insulation which encloses all metal parts.

If an appliance is claimed to be double insulated it should bear the British Standards Institution mark or the certification mark of the British Electrical Approvals Board.

The following are among domestic possibly double insulated appliances: clocks, dishwashers, blanket control units, some heaters (not radiant types), hair driers, power tools, if operated by a motor, shavers and vacuum cleaners.

Q. *How is a consumer unit fixed to a wall?*

A. If the unit has no back, fix a mounting plate on the wall first. This should be metal, insulating material or some other non-combustible material.

Enclosed types can be fitted direct to the wall. Before fitting this type, remove the required knockouts.

Q *What is a splitter unit?*

A. This is a two-way consumer unit with a main switch.

Q. *As socket outlets are not allowed to be fixed in a bathroom, does this mean a ring circuit cannot be run into that room?*

A. The only circumstance in which a ring circuit can be run into a bathroom is when it is needed to supply a fixed appliance such as a wall fire. This fire should be fed by a switched fused connection unit positioned near the fire, but out of reach of a person who is in contact with the bath.

Q. *Can three-pin 13 amp socket outlets be substituted for 15 amp types?*

A. As a rule, yes, but make sure the earthing arrangements are satisfactory. The number of sockets you can connect to each circuit will depend on the size of the cable. This type of outlet will, of course, be wired in the old imperial size cable.

If three-strand cable has been used, only one 13 amp outlet can be connected in place of the old socket outlet. If, however, the circuit is wired in seven-strand cable, more than one outlet can be added, but the value of the circuit fuse will have to be increased.

The new circuit will be known as a radial final sub-circuit to serve only one room. The room involved should not be a kitchen.

This circuit, if fitted with a 20 amp fuse, can supply six socket outlets. The appliances supplied by the outlets should not exceed 13 amps and must not include water heaters.

Q. *I have moved to a brand new house and am told by a friend who understands wiring that the house has been wired strictly in accordance with the regulations. The switches have an earthing conductor and the lighting is wired in the loop-in system.*

I appreciate that to wire up wall lights from the lighting circuit of an old installation might prove complicated, but as my system is modern, would it be a difficult job for me to wire up two wall lights in the lounge using the lighting circuit as a source of supply?

These lights would be in addition to the existing centre light and the wall lights will have built-in switches. I understand the principles of a lighting circuit.

A. There should be no problem in this case. Difficulties can arise in old installations because some of them don't follow recognised patterns and have been added to over the years. Identifying cables in these cases can be difficult, especially to the inexpert.

First you should fit a 30 amp joint box with three terminals to the side of a joist near the centre light.

Take your live and neutral feeds from the switch return and neutral terminals of the ceiling rose to the outer terminals of the joint box. Use 1·0 mm.^{2} twin pvc sheathed cable (no earth) for this purpose.

Then connect a green insulated single core cable from the earthing terminal on the rose to the middle terminal of the joint box.

Where there is no earthing terminal at the rose, this earthing continuity conductor will have to be run right back to the earthing block on the consumer unit.

The cables from the joint box to each wall light should be 1·0 mm.^{2} or 1·5 mm.^{2} twin and earth pvc sheathed.

The existing centre light switch at the door will act as a master switch for the wall lights also. If the centre light is to be *off* when you want the wall lights *on*, you will need to fit a switched lampholder to the flex of the centre light.

Q. *Is it permissible to run an extractor fan off the ring circuit? To reach the nearest socket outlet in the kitchen, the flexible lead would have to be trailed across the sink. Can I avoid this by, perhaps, running a spur from the ring circuit?*

A. I suggest you run a spur cable from the back of the nearest socket outlet. You need to run a two-core and earth pvc sheathed cable, size 2·5 mm.^{2} from the outlet to a fused connection unit. If you choose a switched type of unit, this can act as a master switch for the fan.

Take the flexible lead from the fan to the terminals marked LOAD on the connection unit. The cable from the socket outlet goes to the terminals marked MAINS. Fit a 3 amp fuse to the fused connection unit.

Q. *Does fuse wire deteriorate with age?*

A. Yes. It will deteriorate more rapidly, though, if the current flowing through the circuit is higher than the rating of the fuse.

Q. *Is it safe to use an extension cable to reach short distances while it is still rolled on the drum?*

A. It is wiser to unroll the cable from the drum as there is always a danger that it may become overheated.

Q. *How is the fuel effect achieved on this type of electric heater?*

A. The imitation log or coal unit is illuminated by a coloured bulb (or perhaps more than one). Over the bulb spinners are fixed. Warm air currents are produced by the bulb and set the spinners in motion to give the effect of smoke and flames.

Q. *Will tubular heaters which operate at black heat warm a room satisfactorily?*

A. Much depends on the size of the room, of course. The standard loading of these heaters is 60 watts a foot. Therefore, to get a load of 2,000 watts (equivalent to a two-bar fire) you would need tubes totalling about 33 ft. in length!

The tubes can, of course, be fixed one above the other in banks and they come in lengths up to 17 ft. They are, however, more suitable for background heating.

Q. *What is Electricaire heating?*

A. This is a form of warm air central heating. The warm air is supplied by a special type of storage fan heater which is positioned in a central part of the house.

The heat is ducted through grilles to various rooms.

Q. *How can I tell if my wiring installation needs renewing?*

A. Much depends on the type of cable used. If the insulation and the sheath is rubber, and there is no modern consumer unit installed, it would be wiser to rewire. If the insulation material on any of the conductor ends is brittle and rubs off at a touch, this is a sure indication that a rewire is needed.

If the insulation is pvc, this will not perish with age and should last indefinitely.

Q. *Is it possible to fit a waste disposal unit to a kitchen sink without alteration to the waste hole?*

A. Waste disposal units need a waste hole usually $3\frac{1}{2}$ in. in diameter. If the sink is a metal type, it should be possible to enlarge the hole, but this would not apply to a porcelain sink, of course.

Q. *Is a special circuit needed for door chimes?*

A. No; the wiring circuits for door chimes are similar to those for door bells. However, the method of connecting the wires to the chimes will depend on the make. This should be explained in the instructions supplied with the chimes.

Q. *Is it possible for me to replace the burnt-out element in my electric toaster?*

A. No; this is a job for a service mechanic. To get at the element involves taking the appliance apart. In doing this, the mechanism which causes the bread to pop up when toasted would be disturbed.

Q. *Why is the lead on my television set not connected to the earthing pin of its 13 amp plug?*

A. Normally there are no metal parts exposed on a television set which require earthing. When current is fed to a set it goes to an isolating transformer in the set. This eliminates the need for earthing.

Q. *Can I fit an illuminated bell push to operate a battery powered bell circuit?*

A. It is not advisable to do so because the bulb in the push will be on all the time except when the push is pressed. A battery would be exhausted within a day or so.

The bell should be powered by a double wound mains transformer with an earthed secondary winding if a bell push is an illuminated type.

Q. *When rewiring a first floor lighting circuit, must the cables in the loft pass through holes drilled in the joists, or can they simply lie on the joists?*

A. If the loft has to be boarded over or is to be used for storage purposes, the cables should pass through holes drilled 2 in. below the tops of the joists. Otherwise they should be fixed to the tops of the joists with cable clips. However, as most lofts today have insulation material laid over or between joists, it is wiser, when rewiring, not to leave the cables lying on the joists.

Incidentally, when insulating a loft in this way, it is a good idea to indicate in some way the route the cables take. This, in the event of a future rewire, would avoid the need to move insulation materials unnecessarily to locate the existing cables.

Q. *What are the benefits of fitting twin element immersion heaters?*

A. It is uneconomical to heat a complete cylinder of water if only a small quantity is to be run off. A twin element has, as its name suggests, two elements. One is of sufficient length to reach the base of the hot water cylinder and heat all the water. The other is shorter (perhaps 1 ft. long) and heats just a small amount of water at the top.

Q. *I have seen wall lights connected directly to the terminals of a socket outlet on a ring circuit. Is this the correct way to do this?*

A. No. The wall lights, if operated from the ring circuit, should be fed from a fused connection unit, preferably a switched type, fitted between the socket outlet (or other point on the ring circuit) and the lights. The connection unit should be fitted with a 3 amp fuse.

Q. *What type of wall box is needed to accommodate a fused connection unit?*

A. As the front plate of a connection unit is exactly the same size as an ordinary single socket outlet, the unit can be fitted to a single type steel knockout box. If the connection unit is to be surface mounted, use a plastic socket outlet box. These can be fixed direct to the wall with screws.

Q. *What is the recommended height for socket outlets to be fixed in a kitchen?*

A. The customary height is 3 ft. 6 in. from the ground. Elsewhere, the outlets should be fixed at least 6 in. from the skirting. The higher they are fixed, however, the greater the danger caused by trailing flexes.

Q. *Can socket outlets be used in the garden to supply electric mowers?*

A. If sockets and plugs are used outside they must be adequately protected against the weather. It is better to use weatherproof types.

Q. *I am told that a small storage water heater with a loading of 3kW can be run off a ring circuit. If this is true, why can't an immersion heater also be run off the ring?*

A. Small storage water heaters can be run off the ring circuit (up to 3kW) because they do not operate continuously. An immersion heater is rated as a continuous load.

Q. *What are the four terminals used for in a four-terminal ceiling rose?*

A. Connections for a modern rose with four terminals are: live loop-in; neutral and flex; switch wire and flex; earth.

Q. *Why should the ends of earthing conductors in house cables be enclosed in green insulation sleeving? Those in my wiring installation are left bare.*

A. This is necessary to prevent the earth wire coming into contact with metal parts of an accessory which carry current. This can happen when wiring up a switch or socket outlet. As the outlet or switch front plate is screwed to the wall box, the cable ends are pushed back into the wall box. As this is being done, you have no control over the cables and if a conductor happened to be loose in its terminal it could fly out and make accidental contact as described. If green sleeving were not used, there would be four or perhaps more inches of bare earth wire exposed in the wall box.

Q. *I know my wiring system is out of date and I plan to have it rewired entirely eventually. It works quite well, but I noticed when fitting a dimmer switch in place of an old plate switch that the insulation on the ends of the cable conductors was in a bad way.*
Is there a temporary repair I can make to the insulation.

A. Provided the insulation is faulty *only* at the ends of the conductors, you could insert a short length of red pvc sleeving over the ends, but this should be regarded only as a temporary repair.

Q. *I want to run a couple of wall lights off the ring circuit. Until I redecorate the room, the cables will have to stay on the surface. Is there any way of partially disguising them?*

A. You can get cable cover for this purpose in plastic, metal or wood.

Q. *What is the best position to fit a room thermostat?*

A. Thermostats should not be fitted in the path of a draught. Therefore they should not be fixed near the floor or against the side of a door which opens.
Although thermostats should be freely exposed to air (as opposed to draughts), they should not be directly in the path of radiant heat.
A suitable spot to position them is on an exterior wall about 4ft. 6in. from the level of the floor.

Q. *Are all 13 amp plugs fitted with 13 amp fuses when they are sold in the shops?*

A. This is the common practice. If you want a plug fitted with a 3 amp fuse you could ask the shop people to exchange the 13 amp fitted fuse for the lower rated type. So far as I know, they are not under any obligation to do so, though.

Q. *What are the time switches used for in domestic electric circuits?*

A. Time switches comprise an electric clock which operates ON and OFF contacts at predetermined times. The switches are used for timing off-peak heating circuits, for one thing. Also, plug-in types are available for use in socket outlets. These are for time controlling such things as tea makers and electric blankets. Electric cookers are also fitted with time switches to control the oven.

Q. *What is the difference between an isolating switch and a main switch/fuse unit?*

A. They are both the same. Main switch/fuse units are sometimes referred to as isolating switches because they do, in fact, isolate part of an installation – for example, an installation in an outbuilding supplied with electricity from the house.

Q. *When running a cable on the surface of a wall, at what distance should the cable clips be fastened?*

A. If the cable is run vertically on the wall, the maximum distance between cable clips should be 1ft. 3in. On horizontal runs, the maximum spacing should be 9in.

Q. *Where can I get a copy of the wiring regulations?*

A. The Regulations for the Electrical Equipment of Buildings, to give them their full name, are published by the Institute of Electrical Engineers, Savoy Place, London, W.C.2.

Q. *Instead of a hot water cylinder, we have a rectangular galvanised tank to store the hot water. Can an immersion heater be fitted to it? If so, what is the best position in it for the heater to be fixed?*

A. To ensure that all the water in the tank is heated, fit the immersion heater as low down in the tank as possible, making sure, though, that it does not foul any pipes which may run into the tank.

Index